国家级一流本科专业（网络安全与执法）建设成果
江苏省重点研发计划（江苏警务云安全防护关键技术研究与科技示范）资助
计算机类本科教材

操作系统概念与Linux实践教程

·双语·

杨一涛　编　著
钱发华　卢万进　参　编
赵明生　主　审

電子工業出版社
Publishing House of Electronics Industry
北京 · BEIJING

内 容 简 介

本书按照操作系统的功能分为 10 章，系统地讲述操作系统的基本原理。第 1 章为导论，介绍操作系统的概念、操作系统的形成与发展及操作系统的功能和操作系统提供服务的方式等；第 2～6 章介绍操作系统的进程管理（处理器管理），包括进程、线程、进程调度、进程同步和死锁；第 7 章为内存管理；第 8 章为外存管理；第 9 章为文件管理；第 10 章为 I/O 系统。

为了满足双语教学的需要，本书在每章的最后增加了精选的操作系统技术英文资料，供有能力的读者进行扩展阅读。本书配有 PPT、源代码等教学资源，读者可登录华信教育资源网（www.hxedu.com.cn）免费下载。本书每章都配有教学视频，读者可以扫描书中二维码观看。此外，全书提供 8 个基于 Linux 系统的实验，帮助读者加深对操作系统原理的理解，部分实验也提供演示视频。

本书适合作为本科院校、高职高专院校计算机操作系统和 Linux 相关课程的教材及参考书，也可供相关技术人员参考。

图书在版编目（CIP）数据

操作系统概念与 Linux 实践教程：双语 / 杨一涛编著. 一北京：电子工业出版社，2019.12
ISBN 978-7-121-38197-3

Ⅰ. ①操… Ⅱ. ①杨… Ⅲ. ①操作系统－双语教学－高等学校－教材 ②Linux 操作系统－双语教学－高等学校－教材 Ⅳ. ①TP316

中国版本图书馆 CIP 数据核字(2019)第 296547 号

责任编辑：刘　瑀
印　　刷：北京七彩京通数码快印有限公司
装　　订：北京七彩京通数码快印有限公司
出版发行：电子工业出版社
　　　　　北京市海淀区万寿路 173 信箱　　邮编：100036
开　　本：787×1092　1/16　印张：11.5　字数：294 千字
版　　次：2019 年 12 月第 1 版
印　　次：2024 年 8 月第 3 次印刷
定　　价：39.00 元

凡所购买电子工业出版社图书有缺损问题，请向购买书店调换。若书店售缺，请与本社发行部联系，联系及邮购电话：(010)88254888，88258888。

质量投诉请发邮件至 zlts@phei.com.cn，盗版侵权举报请发邮件至 dbqq@phei.com.cn。

本书咨询联系方式：liuy01@phei.com.cn。

前　言

操作系统是所有计算机系统的重要组成部分。随着计算机设备的多样化和小型化，操作系统也在不断优化。一直以来，操作系统课程是计算机专业的必修课之一，理解操作系统的基本原理是该专业学生学习其他专业课程的必要条件。网络安全与执法专业是公安院校特有的专业方向，专业培养方案中，除公安业务课程外，其他课程与信息安全专业基本一致。因此，操作系统课程也是该专业的核心课程。学习本书前，读者需要掌握以下知识：数据结构、计算机基础及C语言(本书所有代码都是基于C语言编写的)。

在操作系统教材领域，*Operating System Concepts* 一书地位很高，该书作者对操作系统的理解极为深入，写作水平超群。本书以该书为基础进行编写，适当缩减内容，适合48学时或56学时教学使用，希望本书可以用通俗易懂的语言引领读者打开操作系统这扇大门。

操作系统源自西方国家，因为语言和文化差异，很多理论知识必须要读外文原文方可理解透彻，所以本书在每章的最后增加了精选的操作系统技术英文资料，希望使读者在学习过程中得到一些启迪。

为方便读者学习，本书配有 PPT、源代码等教学资源，读者可登录华信教育资源网(www.hxedu.com.cn)下载。本书每章都配有教学视频，读者可以扫描书中二维码观看，也可以在B站搜索本课程或用户“Y4NGY”进行观看。

目前，国产操作系统的发展蒸蒸日上，本书以国产的深度系统(Deepin)为实验平台，通过 8 个实验让读者对操作系统原理有更直观的认识。部分实验配有对应的演示视频，希望这些能给读者的学习带来更多的便利。

本书第1～9章由杨一涛编写，第10章由钱发华、卢万进编写，全书由赵明生教授审阅。本书在编写过程中得到王新猛、吴育宝及吴玉强等老师的关心和支持，为本书配图和调试实验代码的学生有刘玮健、张宇洁、潘宏晨、陈珣瑜、刘俊涛、赵博文、赵博婧、吴王范、曹开焱、田心玥、徐立力、蒋佳宇及童芝丹。在此对所有帮助过我们的人表示衷心的感谢！由于作者水平有限，书中难免有错误之处，有任何建议或疑问可通过邮箱 youngyt@gmail.com 联系。

编著者

目　录

第1章　导　　论

1.1　操作系统的概念

获取视频

1. 什么是操作系统

我们日常生活中经常接触的机器，如洗衣机、烤箱及电冰箱等，其内部的运作原理可能未知，但是这些机器大多都提供了一个“操作面板”，用户可以通过这个面板向机器下达任务指令。以给洗衣机设定30分钟的快洗任务为例，洗衣机在收到任务指令后，就会按照要求工作30分钟。除了“操作面板”，这些机器后面还有一套“软件系统”在支撑机器的运转。它负责将用户在面板上的操作转换成相应的指令，然后将指令传给送目标部件(如滚筒马达)。它能设置一个30分钟的定时器,在定时器时间到时发出信号,使机器停止运转,并收集运转结果(正常结束或出现故障)，将结果通过“操作面板”(如指示灯)反馈给用户。在这个例子中，可以将接收用户操作的“操作面板”及负责下达指令并控制部件的“软件系统”粗略地视为这台机器的操作系统。

2. 计算机操作系统

本书要讲述的对象是计算机操作系统。计算机可以大致分为硬件系统、操作系统、应用程序和用户。现代计算机的硬件系统由一个或多个处理器、主存储器、外存储器、键盘、鼠标、显示器及打印机等输入/输出(I/O)设备组成。这个硬件系统非常复杂，直接对硬件进行操作对用户而言是不切实际的，因此在这些硬件之上需要加装一层软件，即计算机操作系统。该系统是配置在计算机硬件上的第一层软件，是对硬件系统的首次扩充。它在计算机中占据了特别重要的地位，而其他诸如汇编程序、编译程序、数据库管理系统等的系统软件，以及大量的应用软件，都将依赖操作系统，都要取得它的服务。操作系统已成为现代计算机系统、多处理器系统、计算机网络、多媒体系统及嵌入式系统中必须配置的最重要的系统软件。本书所提及的“操作系统”一词不加特殊说明均指计算机操作系统。

3. 操作系统的角色

对于一个完全无软件的计算机(俗称“裸机”)来说，它向用户提供的是实际硬件接口(物理接口)，用户必须对物理接口的实现细节有充分的了解，并利用机器指令进行编程，才能够实现对计算机硬件的操作，因此“裸机”必定是难以使用的。为了方便用户使用I/O设备，人们在“裸机”上增加了一层I/O设备管理软件(简称I/O软件)。

人们用I/O软件来实现对I/O设备的操作，它向上提供一组I/O操作指令，如Read、Write指令，用户可利用这些指令进行数据的输入或输出，而无须关心I/O操作是如何实现的。此时用户所看到的机器将是一台比“裸机”功能更强、使用更方便的机器。也就是说，在“裸机”上铺设的I/O软件隐藏了I/O操作的细节，向上提供了一组抽象的I/O设备，如图1-1所示。

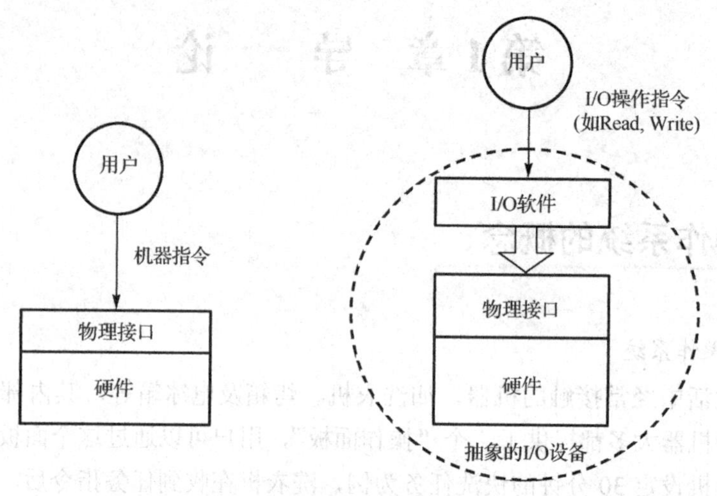

图1-1 I/O软件隐藏I/O操作的细节

操作系统扮演的角色就是这组抽象的I/O设备，它向用户提供了一个对硬件操作的抽象模型。用户可利用抽象模型提供的接口使用计算机，而无须了解物理接口实现的细节，从而使用户更容易地使用计算机的硬件。我们通常称操作系统为“虚拟计算机”，它向用户提供友好的使用界面。用户向这台“虚拟计算机”发出指令是通过鼠标及键盘等完成的，而“虚拟计算机”把用户抽象的指令转换成了对计算机硬件的真实操作。除了操作硬件，操作系统提供了额外的软件(文件系统)帮助用户方便地管理文档。同样地，它实现了对文件的操作，并向用户提供一组对文件进行存取操作的接口。用户可利用这组接口进行文件的存取(如Windows系统的“资源管理器”和macOS系统的“访达”)。当用户有新的功能需求时，设计人员会不断地对操作系统进行扩充，而这台“虚拟计算机”就变得越来越强大，操作也就越来越方便。

由此可知，操作系统是铺设在计算机硬件上的多层系统软件。它们不仅增强了系统的功能，而且隐藏了对硬件操作的细节，由它们实现了对计算机硬件操作的多个层次的抽象。值得说明的是，对一个硬件在底层进行抽象后，在高层还可再次对其进行抽象，使其成为更高层的抽象模型。随着抽象层次的提高，抽象接口所提供的功能越来越强，用户使用起来也就越来越方便。

4. 操作系统的定义

计算机的历史虽短，但发展迅猛，随着计算机功能的不断增强和体积的不断减小，大量不同的操作系统出现。一般来说，我们无法对操作系统进行完全准确的定义。操作系统的存在是因为它提供了合理的方式来解决创建可用计算机系统的问题，而计算机系统的根本目的是使用户程序能更容易地解决用户问题。为了实现这一目的，人们构造了计算机硬件，但由于计算机硬件本身不容易使用，因此人们又开发了应用程序。这些应用程序需要一些通用功能，如控制各种I/O设备及共享硬件资源。这些通用功能被组装成了一个软件模块，即操作系统(Operating System，OS)。

1.2 操作系统的形成与发展

1.2.1 手工操作阶段

第一代计算机(从1946年到20世纪50年代中期)是利用成千上万个真空管做成的，它体积庞大且运行速度仅为数千次每秒，功耗也非常高。这时，还未出现软件形态的操作系统，用户采用人工操作方式直接使用计算机硬件系统，将事先已穿孔(对应于程序和数据)的纸带装入纸带输入机，将纸带上保存的程序代码和数据输入计算机，然后启动计算机。当程序运行完毕时，将计算结果输出到纸带上之后，才让下一个用户使用计算机。

这种人工操作方式有以下两方面缺点。

(1)用户独占机器。计算机及全部资源只能由上机用户独占。

(2)CPU等待人工操作。当用户进行装带及卸带等人工操作时，CPU及内存等资源是空闲的。

可见，人工操作方式严重降低了计算机资源的利用率，从而产生人机矛盾。随着CPU速度的提高及系统规模的扩大，人机矛盾日趋严重。此外，CPU的速度迅速提高而I/O设备的速度却提高得很缓慢，这又使CPU与I/O设备之间速度不匹配的矛盾更加突出。

1.2.2 单道批处理系统

第二代计算机从20世纪50年代开始出现，人们开始用体积更小、功耗更低的晶体管替代真空管来制造计算机。为了能充分利用计算机，尽量让计算机系统连续运行，以减少空闲时间，通常把一批作业(Job)以脱机(Offline)的方式输入磁带，并在系统中事先启动一个监督程序(Monitor)，负责控制这批作业一个接一个地被连续处理。其自动处理过程如下：首先，由监督程序将磁带上的第一个作业装入内存，并把运行控制权交给该作业；当该作业处理完成时，又将控制权交还给监督程序，再由监督程序把磁带上的第二个作业装入内存；计算机系统就这样自动地一个接一个地对作业进行处理，直至磁带上所有的作业全部完成。这个监督程序就是操作系统的雏形，被称为“批处理系统”。图1-2给出了单道批处理系统(Simple Batch Processing System)的处理流程，这里的“单道”是指内存中始终保持只有一个作业。

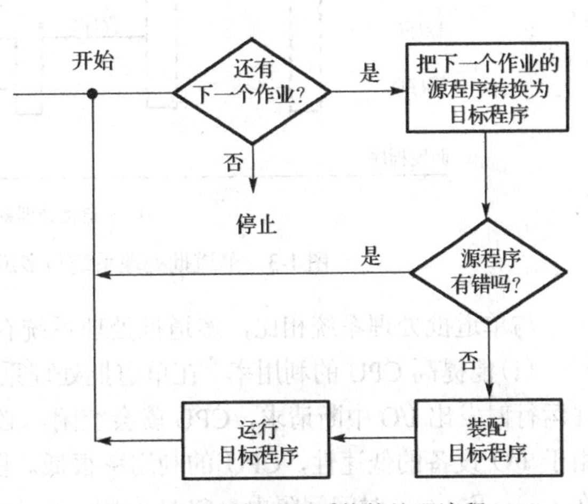

图1-2 单道批处理系统的处理流程

不难看出，单道批处理系统的出现在一定程度上解决了人机矛盾，多个用户的计算任务可以集中被批量处理，换句话说，提高了整个系统的资源利用率和系统吞吐量。但是，I/O设备相对CPU而言，速度还慢得多，

I/O 设备在工作时，CPU 处于空闲状态，二者速度不匹配的问题并没有得到很好的解决。于是，人们开始设计一种允许多道作业同时驻留内存的系统，即“多道批处理系统”。

1.2.3　多道批处理系统

在单道批处理系统中，内存中仅有一个作业，它无法充分利用系统中的所有资源，致使系统性能较差。为了进一步提高资源利用率和系统吞吐量，在 20 世纪 60 年代中期，又出现了多道程序设计(Multi-programming)技术，由此形成了多道批处理系统(Multi-programmed Batch Processing System)。在该系统中，用户所提交的作业都先存放在外存上，并排成一个队列，称为“后备队列”；然后，由作业调度程序(Scheduler)按一定的算法从后备队列中选择若干作业调入内存，使它们共享 CPU 及系统中的各种资源，特别是 CPU 资源。它的中心思想是,如果内存中一个正在运行的作业启动了 I/O 设备，调度程序则会调度内存中的另一个作业，使用 CPU 进行计算。这样，CPU 始终处于忙碌状态，利用率得到很大的提升。图 1-3 给出了单道批处理系统与多道批处理系统运行程序的区别。

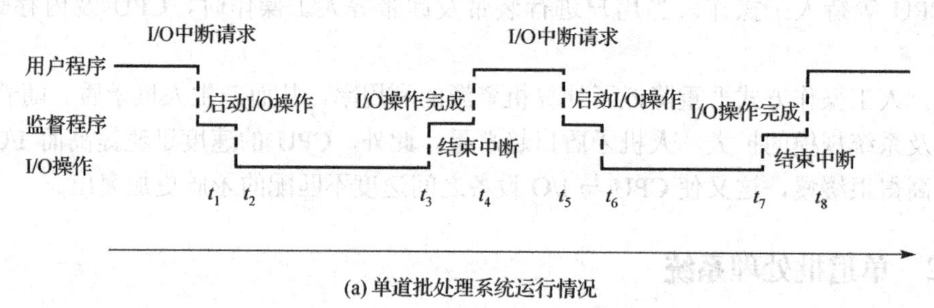

(a) 单道批处理系统运行情况

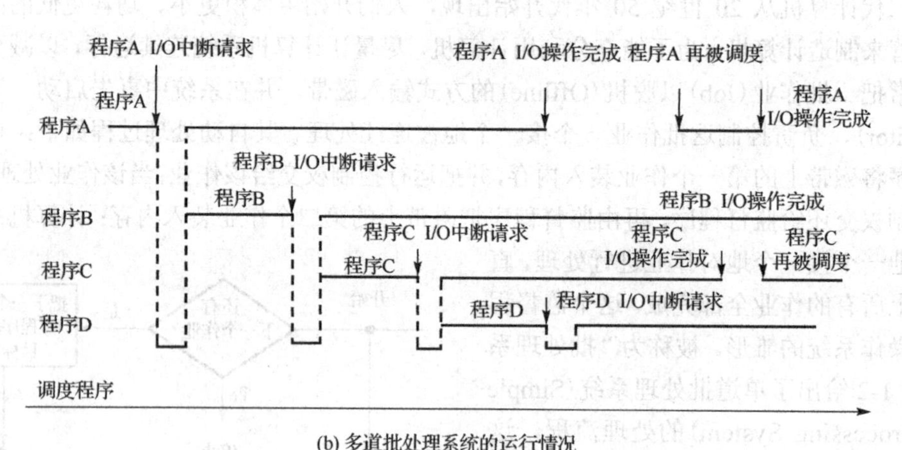

(b) 多道批处理系统的运行情况

图 1-3　单道批处理系统与多道批处理系统运行程序的区别

与单道批处理系统相比，多道批处理系统有以下特征。

(1) 能提高 CPU 的利用率。在单道批处理系统中，当内存中仅有一个程序时，每逢该程序在运行时发出 I/O 中断请求，CPU 就会空闲，必须在其 I/O 操作完成后，CPU 才能继续运行；由于 I/O 设备的低速性，CPU 的利用率很低。图 1-3(a)给出了单道批处理系统的运行情况，在 t_2～t_3 及 t_6～t_7 时间间隔内，CPU 空闲。引入多道批处理系统后，由于在内存中同时装有若干程序，并使它们交替运行，因此当正在运行的程序因 I/O 中断请求而暂停运行时，系统可调度其他程序运行，从而使 CPU 保持忙碌状态。图 1-3(b)给出了多道批处理系统中四个程序的运行情况。

(2)可提高内存和 I/O 设备的利用率。通常来说，内存容量相对作业大小而言要大得多，由于 80%左右的作业都属于中小型作业，因此在单道批处理系统中，同一时间，内存只驻留一个作业会造成内存空间的浪费。如果允许同时驻留多个作业，那么这些作业可以在调度程序的控制下有序地交替运行，我们把这种现象称为“并发”(Concurrency)。多个作业的并发运行无疑大大提高了内存及 I/O 设备的利用率。

(3)增加系统吞吐量。在保持 CPU、I/O 设备忙碌状态的同时，必然会大幅度增加系统的吞吐量，从而减少作业加工所需的费用。

多道批处理系统虽然具备很多优势，但无疑增加了设计的复杂性，为了使系统中的多个程序能够协调运行，会有一系列的软硬件问题需要解决，随着以下技术的出现，多道批处理系统被不断地完善。

(1)通道技术。该技术提供了 CPU 与 I/O 设备的并行能力，使得 CPU 和 I/O 设备可以独立地同时运行，即当某个作业占用 CPU 时，其他作业可以执行 I/O 操作。

(2)中断(Interrupt)技术。多个程序运行的特点是，作业不会一直占用 CPU，当它执行 I/O 操作时会暂时让出 CPU，等待 I/O 操作结束后，它才会被再次调入 CPU。中断技术保证了作业在重新回到 CPU 时，可以接着之前未完成的操作继续运行，而不是重新开始。

(3)大容量存储技术。内存容量的增加，允许更多的作业同时驻留在内存中等待调度；大容量存储设备(磁盘)的问世催生出了“交换”(Swapping)技术，它实现了用外存储器充当虚拟内存的功能。在理论上，计算机的内存是“无限”大的。

1.2.4 分时操作系统

在计算资源比较匮乏的时代，为了让多个用户可以共享一台计算机，分时操作系统(Time Sharing Operating System)问世了。每个用户都拥有一个终端(包括显示设备和输入设备)，所有用户终端均与一台计算机相连，分时操作系统给所有用户轮流分配一个时间片(Time Slice)，用户终端只有在得到时间片的时候才可使用计算资源运算，时间片一旦用完，就会被强制下线，等待下一轮时间片的分配。该系统创造性地提出了一种多任务并发控制机制，只要时间片大小和用户终端数量控制合理，所有用户就感觉不到他们在共享一台计算机，仿佛每个人都独占着计算资源。

第一个真正的分时操作系统是由麻省理工学院开发成功的兼容分时系统(Compatible Time Sharing System，CTSS)。继 CTSS 成功后，麻省理工学院又和贝尔实验室及通用电气公司联合开发出多用户多任务操作系统——MULTICS，其能支持数百个用户。值得一提的是，参加 MULTICS 研发的贝尔实验室专家 Ken Thompson，在 PDP-7 小型机上开发出了一个简化的 MULTICS 版本，它就是当今广为流行的 UNIX 操作系统的前身。

1.2.5 微机操作系统

配置在微机(微型机)上的操作系统称为微机操作系统，最早诞生的微机操作系统是配置在 8 位微机上的 CP/M。当微机发展为 16 位、32 位、64 位时，相应的操作系统也应运而生。

1. 单用户单任务操作系统

单用户单任务操作系统的含义是，只允许一个用户上机，且只允许用户程序作为一个任务运行。这是最简单的微机操作系统，主要配置在8位和16位微机上。最有代表性的单用户单任务微机操作系统是CP/M及MS-DOS。

(1)CP/M。1974年，在第一代通用8位微处理器芯片Intel 8080出现后的第二年，Digital Research公司就开发出带有软盘系统的8位微机操作系统。1977年，Digital Research公司对CP/M进行了重写，使其可配置在以Intel 8080、8085及Z80等8位芯片为基础的多种微机上。1979年，该公司又推出带有磁盘管理功能的CP/M 2.2版本。

(2)MS-DOS。1981年，IBM公司首次推出了IBM-PC个人计算机(16位微机)，在微机中采用了微软公司开发的MS-DOS (Disk Operating System)操作系统，该操作系统在CP/M的基础上进行了较大的扩充，功能有很大的增强。1983年，IBM推出了配有Intel 80286芯片的PC/AT计算机，与之相应的是MS-DOS 2.0版本，它不仅能支持磁盘设备，还采用了树形目录结构的文件系统。从1989年到1993年，有多个MS-DOS版本推出，它们都可以安装在Intel 80386、80486等32位微机上。从20世纪80年代到90年代，MS-DOS成为16位单用户单任务操作系统的标准。

2. 单用户多任务操作系统

单用户多任务操作系统同样只允许单个用户登录，但允许用户把程序分为若干任务，使它们并发运行，从而有效地改善了系统的性能。目前在32位微机上配置的操作系统基本上都是单用户多任务操作系统，其中最有代表性的是微软推出的Windows系统。1985年及1987年，微软先后推出了Windows 1.0及Windows 2.0系统。1990年，微软又发布了Windows 3.0系统，随后又发布了Windows 3.1系统，它们主要是针对Intel 8038及80486等32位微机开发的，较以前的操作系统有很大的改进，引入了友好的图形用户界面，支持多任务和扩展内存的功能，从而成为Intel 80386及80486等微机的主流操作系统。1995年，微软推出了Windows 95系统，它较之以前的Windows 3.1系统有许多重大改进，采用了全32位处理技术，并兼容16位应用程序，同时，还集成了支持Internet的网络功能。1998年，微软又推出了Windows 95的改进版Windows 98，它是最后一个仍然兼容16位应用程序的Windows系统，其最主要的改进是把微软自己开发的Internet浏览器整合到操作系统中，大大方便了用户上网，此外，它还增加了对多媒体的支持。2001年，微软又发布了Windows XP系统，同时提供了家用和商业工作站两个版本。2009年以来，Windows 7、Windows 8、Windows 10陆续发布。

3. 多用户多任务操作系统

多用户多任务操作系统含义是，允许多个用户通过各自的终端使用同一台机器，共享主机系统中的各种资源，而每个用户程序又可进一步分为几个任务，它们能并发运行，从而可进一步提高资源利用率和系统吞吐量。在大、中及小型机中所配置的大多是多用户多任务操作系统，其中最有代表性的是UNIX操作系统，它是美国电话电报公司的贝尔实验室在1969—1970年期间开发的，1979年推出的UNIX v7已被广泛应用于多种中、小型机上。后来基于UNIX出现了很多变种，其中“最闪耀的明星”自然是开源的操作系统GNU/Linux，它的内核最初由芬兰学生Linus Torvalds开发并于1991年连同源代码在Internet上发布。

1.2.6 分布式操作系统

分布式操作系统负责管理分布式处理系统资源及控制分布式程序运行。它和集中式操作系统的区别体现在资源管理、进程通信及系统结构等方面。

分布式程序设计语言用于编写运行于分布式处理系统上的分布式程序。一个分布式程序由若干可以独立运行的程序模块组成，它们分布于一个分布式处理系统的多台计算机上同时运行。分布式程序与集中式程序相比有三个特点：分布性、通信性及稳健性。

分布式文件系统具有远程文件存取的功能，并以透明方式对分布在网络上的文件进行管理和存取。

分布式数据库系统由分布于多个计算机节点上的若干数据库系统组成，它提供有效的存取手段来操纵这些节点上的子数据库。分布式数据库在使用上可视为一个完整的数据库，而实际上是分布在地理上分散的各个节点上的。当然，分布在各个节点上的子数据库在逻辑上是相关的。分布式数据库系统由若干站集合而成，这些站又称为节点，它们在通信网络中连接在一起，每个节点都是一个独立的数据库系统，都拥有各自的数据库、中央处理器(CPU)、终端及局部数据库管理系统。因此分布式数据库系统可以视为一系列集中式数据库系统的集合。它们在逻辑上属于同一系统，但在物理结构上是分布式的。

1.2.7 嵌入式操作系统

嵌入式操作系统(Embedded Operating System，EOS)是指用于嵌入式系统的操作系统，嵌入式系统包括但不限于微波炉、电视机、手机、汽车等。嵌入式操作系统通常包括与硬件相关的底层驱动软件、系统内核、设备驱动接口、通信协议、图形用户界面及标准化浏览器等。目前，在嵌入式领域广泛使用的操作系统有嵌入式 Linux、QNX 和 VxWorks，以及应用在智能手机和平板计算机中的 Android、iOS 等。

1.3 操作系统的功能

1.3.1 接管计算机

磁盘存储器不仅容量大、存取速度快，而且可以实现随机存取，是当前存放大量程序和数据的理想设备，因此在现代计算机系统中，都配置了磁盘存储器，主要用于存放文件。磁盘设备可包括一个或多个物理盘片，每个盘片分为一个或两个存储面，也称盘面(Surface)，如图 1-4(a)所示，每个盘面被组织成若干同心环，这种环称为磁道(Track)，各磁道之间留有必要的间隙。每条磁道又在逻辑上被划分成若干扇区(Sector)，扇区也称盘块、物理块(Block)，各扇区之间需要保留一定的间隔。图 1-4(b)中共有三条磁道，每条磁道被个被划分出了多个扇区，若规定每个扇区存储的字节数是相同的，那么外圈磁道的存储密度就比内圈磁道的存储密度要低。

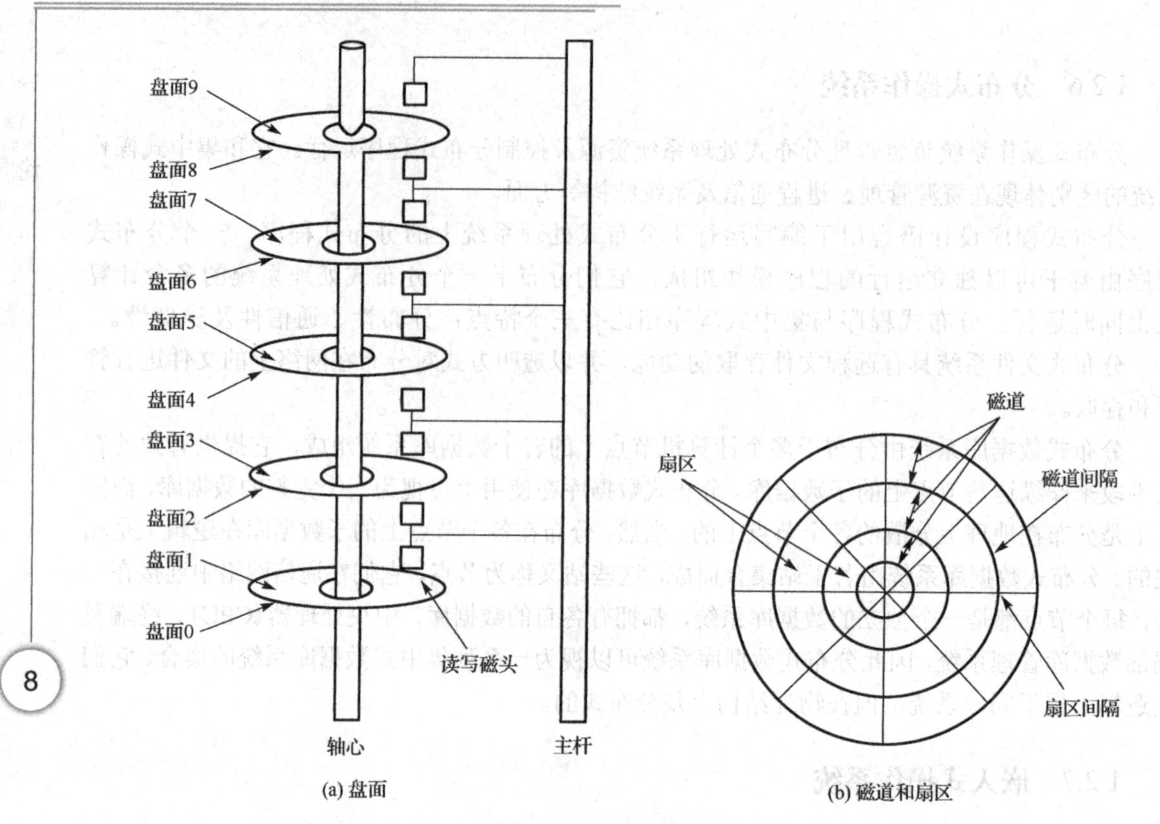

图 1-4　磁盘的结构和布局

一条物理记录存储在一个扇区上，磁盘上存储的物理记录的条数是由扇区数、磁道数及盘面数决定的。例如，一个 10 GB 容量的磁盘，有 8 个双面可存储的盘片，共 16 个盘面，每面有 16383 条磁道，63 个扇区。

当计算机电源被打开时，会先进行加电自检，然后寻找启动盘，如果计算机选择从软盘启动，就会检查软盘的 0 面 0 磁道 1 扇区，如果发现它以 0xaa55 结束，那么 BIOS(Basic Input Output System，基本输入输出系统)就认为它是一个引导扇区。当然，引导扇区的结构我们会在下文中介绍。一旦 BIOS 发现了引导扇区，它就会将这 512 字节的内容，装载到内存的 0000:7c00 处，然后跳转到该位置处将控制权彻底交给这段引导代码。到此为止，计算机不再由 BIOS 中固有的程序来控制，转由操作系统的一部分来控制。

从打开电源到开始操作，计算机的启动是一个非常复杂的过程。“启动”在英语中用 boot 来表达，它来自一句谚语：“Pull oneself up by one’s bootstraps”，字面意思是“拽着鞋带把自己拉起来”。工程师们最初用它来比喻计算机启动是一个很矛盾的过程：必须先运行程序，然后计算机才能启动，但是计算机不启动就无法运行程序。于是他们想尽各种办法，把一小段程序先装进内存，然后把计算机“拉”启动，那一小段程序就是“鞋带”，整个启动过程就简称为 boot 了。计算机的整个启动过程可分成四个阶段。

1. 第一阶段：BIOS

20 世纪 70 年代初，只读存储器(Read-Only Memory，ROM)出现，开机程序被刷入 ROM 中，计算机通电后，第一件事就是读取开机程序，该程序即 BIOS。

1)硬件自检

BIOS 程序首先检查计算机硬件能否满足运行的基本条件，这称为“硬件自检”(Power-On Self Test，POST)。如果硬件出现问题，主板会发出不同含义的蜂鸣声，启动终止。如果没有问题，屏幕就会显示出 CPU、内存及磁盘等信息，如图 1-5 所示。

```
Diskette Drive B  : None                     Serial Port(s)     : 3F0 2F0
Pri. Master Disk  : LBA,ATA 100,  250GB Parallel Port(s)  : 370
Pri. Slave  Disk  : LBA,ATA 100,  250GB DDR at Bank(s)    : 0 1 2
Sec. Master Disk  : None
Sec. Slave  Disk  : None

Pri. Master Disk  HDD S.M.A.R.T. capability ... Disabled
Pri. Slave  Disk  HDD S.M.A.R.T. capability ... Disabled

PCI Devices Listing ...
Bus  Dev  Fun  Vendor Device  SVID  SSID Class  Device Class              IRQ

 0   27    0   8086   2668    1458  A005  0403  Multimedia Device          5
 0   29    0   8086   2658    1458  2658  0C03  USB 1.1 Host Cntrlr        9
 0   29    1   8086   2659    1458  2659  0C03  USB 1.1 Host Cntrlr       11
 0   29    2   8086   265A    1458  265A  0C03  USB 1.1 Host Cntrlr       11
 0   29    3   8086   265B    1458  265A  0C03  USB 1.1 Host Cntrlr        5
 0   29    7   8086   265C    1458  5006  0C03  USB 1.1 Host Cntrlr        9
 0   31    2   8086   2651    1458  2651  0101  IDE Cntrlr                14
 0   31    3   8086   266A    1458  266A  0C05  SMBus Cntrlr              11
 1    0    0   10DE   0421    10DE  0479  0300  Display Cntrlr             5
 2    0    0   1283   8212    0000  0000  0180  Mass Storage Cntrlr       10
 2    5    0   11AB   4320    1458  E000  0200  Network Cntrlr            12
                                                ACPI Controller            9
```

图 1-5　计算机启动界面信息

2)启动顺序

硬件自检完成后，BIOS 把控制权转交给下一阶段的启动程序。这时，BIOS 需要知道，“下一阶段的启动程序”具体存放在哪个设备中。也就是说，BIOS 需要知道外部存储设备的顺序，排在前面的设备就是优先转交控制权的设备。这种顺序称为“启动顺序”(Boot Sequence)。打开 BIOS 操作界面，里面有一项就是“设定启动顺序”。

2. 第二阶段：主引导记录

BIOS 按照“启动顺序”把控制权转交给排在第一位的存储设备。这时，计算机读取该设备的第一个扇区(启动扇区)，也就是读取最前面的 512 字节。如果这 512 字节的最后两字节是 0x55 和 0xaa，表明这个设备可以用于启动；如果不是，表明设备不能用于启动，控制权被转交给“启动顺序”中的下一个设备。这最前面的 512 字节，称为“主引导记录”(Master Boot Record，MBR)。

3. 第三阶段：磁盘

这时，计算机的控制权就要转交给磁盘的某个分区了，这里又分成三种情况。

1)卷引导记录

计算机会读取激活分区的第一个扇区，称为“卷引导记录”(Volume Boot Record，VBR)。“卷引导记录”的主要作用是告诉计算机操作系统在这个分区中的位置，然后计算机就会加载操作系统了。

2)扩展分区和逻辑分区

随着磁盘越来越大，需要更多的分区。但是，分区表中只有四项，因此规定有且仅有一个分区可以被定义成“扩展分区”(Extended Partition)。

“扩展分区”是指这个分区又可分成多个分区。这种分区里面的分区，称为“逻辑分区”

(Logical Partition)，扩展分区可以包含无数个逻辑分区。但是，很少通过这种方式启动操作系统。

3)启动管理器

在这种情况下，计算机读取“主引导记录”前面 446 字节的机器码之后，不再把控制权转交给某一个分区，而是运行事先安装的“启动管理器”(Boot Loader)，由用户选择启动哪个操作系统。在 Linux 环境中，目前最流行的启动管理器是 GRUB，如图 1-6 所示。

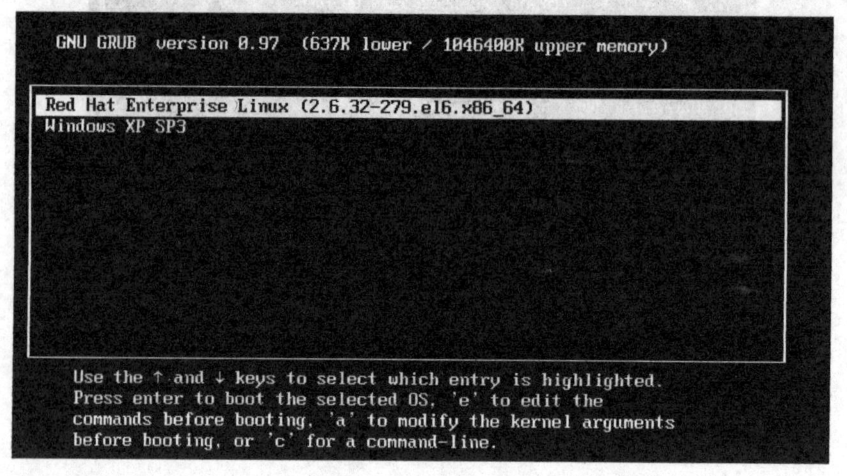

图 1-6 GRUB 界面

4. 第四阶段：操作系统

控制权被转交给操作系统后，操作系统的内核首先被载入内存。以 Linux 系统为例，先载入/boot 目录下面的 kernel。内核加载成功后，第一个运行的程序是/sbin/init。它根据配置文件产生 init 进程。这是 Linux 系统启动后的第一个进程，进程编号(pid)为 1，其他进程都是它的后代。然后，init 进程加载系统的各个模块，如窗口程序和网络程序，直至运行/bin/login 程序，出现登录界面，等待用户输入用户名和密码。

至此，全部启动过程完成。

1.3.2 进程管理

在计算机领域，“进程”(Process)一词用来描述程序运行之后的形态，“线程”(Thread)是一种比进程粒度更小的形态，而“程序”(Program)特指外存储器中的二进制可执行文件。在进入第 2 章之前我们暂时不去考虑它们的区别，因为一些大型的软件在运行时不止调用一个程序，所以在操作系统中，总将软件的运行抽象成一个作业的运行。

在多道批处理系统中，处理器的分配和运行都以进程为基本单位，因此对处理器的管理可归结为对进程的管理。在引入了线程的操作系统中，也包含对线程的管理。处理器管理的主要功能是创建和撤销进程(线程)，对各进程(线程)的运行进行调度，实现进程(线程)之间的信息交换，以及按照一定的算法把处理器分配给进程(线程)。

在多道批处理环境下，要使作业运行，必须先为其创建一个或几个进程，并为其分配必要的资源(包括计算资源及信息资源)。当进程运行结束时，立即撤销该进程，并及时回收该进程所占用的各类资源。引入线程的操作系统还应具备创建及撤销线程的功能。

多道并发的进程是以异步(随机)方式运行的，并以不可预知的速度向前推进。为使多个进程能有条不紊地运行，系统中必须设置进程同步机制。进程同步机制的主要任务是对多个进程的运行进行协调，以保证无论进程的运行顺序如何，结果都是唯一且正确的。进程同步方式主要有两种。

(1)进程互斥方式。是指进程在竞争临界资源(是指同一时刻只能被一个进程占用的资源，如打印机)时，应采用互斥方式。操作系统中最常用的互斥工具是“锁”，操作系统为每个临界资源配置一把锁，当锁打开时，进程可以对该临界资源进行访问；而当锁闭合时，则禁止进程访问该临界资源。

(2)进程同步方式。是指多个进程在相互合作去完成共同任务时，用同步机制对它们的运行次序加以协调。为了实现进程同步，系统中必须设置进程同步机制，常见的进程同步机制有信号量与 P、V 操作[P 表示 Passeren(通过)，V 表示 Vrijgen(释放)]及“管道”等。实际上，互斥是一种特殊的同步，即进程交替使用资源。

以上进程同步方式同样适用于线程。

1.3.3 存储管理

操作系统存储管理的主要任务是为多道程序的运行提供良好的环境，方便用户使用存储器，提高存储器利用率，以及能从逻辑上扩充内存。因此，存储管理应包括内存分配、内存保护、地址映射及内存扩充等功能。

1. 内存分配

内存分配的主要任务是为每道程序分配供其运行的内存空间；通过优化分配策略以提高存储器的利用率，减少不可用的内存碎片的数量；对于正在运行的程序，允许其根据运行需要动态地申请新的内存空间。

操作系统在实现内存分配时，一般采取静态及动态两种方式。在静态分配方式中，每个程序的内存空间是在装入前确定的，在程序装入内存后的整个运行期间，不允许再申请新的内存空间；在动态分配方式中，操作系统允许程序在运行过程中申请新的附加内存空间，以适应程序和数据的动态增长，也允许其在内存中“移动”。

2. 内存保护

内存保护的主要任务是确保每个用户程序都只在自己的内存空间中运行，彼此互不干扰；绝不允许用户程序访问操作系统的代码及数据；也不允许用户程序跳转到非共享的其他用户程序内存空间中去运行。一种比较简单的内存保护机制是设置两个界限寄存器，分别用于存放正在运行程序的上界及下界，系统对每条指令所要访问的地址进行检查，若地址超出两个边界，即发生越界现象，则发出越界中断请求，以停止该程序的运行。一般越界检查都由硬件完成，可以加快程序的运行速度。

3. 地址映射

应用程序(源程序)经编译、链接后，会生成可执行二进制指令列表，每条二进制指令都有一个数字形式的编号，每个程序的二进制指令列表总是从“0”开始编号的，后续指令编号都是相对于起始地址计算的，这些指令编号称为“逻辑地址”或“相对地址”，形成的地址范围称为“地址空间”。此外，由内存中的一系列单元限定的地址范围称为“内存空间”，其中的地址称为“物理地址”。程序的逻辑地址不可能都和物理地址一致，为使程序正确运行，存储管理必须提供地址映射功能，以将地址空间中的逻辑地址转换为内存空间中与之对应的物理地址。

4. 内存扩充

存储管理中的内存扩充任务并非去扩大物理内存的容量，而是借助虚拟存储技术，从逻辑上去扩充内存容量，使用户感觉到的内存容量比实际内存容量大得多，以便让更多的用户程序并发运行。这样，既能满足用户的需要，又能改善系统的性能。为此，只需增加少量的硬件。为了能在逻辑上扩充内存，系统必须具有内存扩充机制，用于实现下述功能。

(1) 请求调入功能。系统允许在仅装入一部分用户程序和数据的情况下启动该程序。在程序运行过程中，若发现要继续运行时所需的程序及数据尚未装入内存，可向操作系统发出请求，由操作系统从磁盘中将所需部分调入内存，以便继续运行。

(2) 置换功能。若发现在内存中已无足够的空间来装入需要调入的程序及数据，系统应能将内存中的一部分暂时不用的程序及数据调至磁盘上，以腾出内存空间，然后再将所需的部分装入内存。

1.3.4 文件管理

在现代计算机管理中，程序及数据是以文件的形式存储在磁盘上，供所有或指定的用户使用的。操作系统有一个专门的文件子系统负责文件管理功能，它具备文件存储空间管理、目录管理、文件读/写管理及保护等功能，主要任务是对用户文件和系统文件进行管理，以方便用户使用，并保证文件的安全性。

1. 文件存储空间管理

文件一般存储在外存储器（简称外存）中，文件系统首先要考虑的问题是如何划分存储空间进行文件存储，特别是外存容量特别大的情况。此外，在多用户环境下，如何提高多用户存储管理文件的效率也是文件系统需要解决的问题。

2. 目录管理

文件是按名存取的，这个“名”就是我们熟知的文件名。为了避免二义性，文件名是不允许重复的，但是考虑到多用户命名文件的灵活性，早期的文件系统开始引入“目录”的概念。目录也就是我们常说的文件夹，每个用户都拥有一个以自己用户名命名的目录，在用户各自目录中命名的文件名互相不受影响，可以重名。随后目录结构被扩展，目录还可以创建子目录，子目录下可以创建下一级子目录，最终形成我们现在多数文件系统采用的树形目录结构。

目录除了可以将文件进行归类管理，它最重要的功能是文件索引，目录能使用户方便地在外存上找到自己所需的文件。目录由若干目录项构成，每个目录项包括文件名、文件属性及文件在磁盘上的物理位置等。用户在读/写文件时，首先要提供文件名，然后按文件名在目录中进行检索，如果可以找到该目录项，则可以依据目录项中所记录的磁盘位置加载文件。目录管理还能提供快速的目录查询手段，以提高对文件的检索速度。

3. 文件读/写管理及保护

1) 文件读/写管理

该功能能够实现根据用户的请求，从外存中读取数据，或将数据写入外存。在进行文件读/写时，系统先根据用户给出的文件名去检索目录，从中获得文件在外存中的位置。然后，利用文件读/写指针，对文件进行读/写。一旦读/写完成，便修改读/写指针，为下一次读/写做好准备。由于读/写操作不会同时进行，因此可合用一个读/写指针。

2) 文件保护

为了防止系统中的文件被非法窃取和破坏，在文件系统中必须提供有效的保护功能，以实现下述目标：防止未经核准的用户存取文件；防止冒名顶替的用户存取文件；防止用户以不正确的方式使用文件。

1.3.5 设备管理

计算机的外围设备(Peripheral Device)也称为 I/O 设备，它们种类繁多、功能各异，而且新式的设备不断涌现，操作系统的编写人员进行了巧妙的设计来应对日新月异的设备管理问题。操作系统 I/O 子系统的主要功能就是管理所有 I/O 设备，主要任务包括处理用户进程提出的 I/O 中断请求，为用户进程分配其所需的 I/O 设备，提高 CPU 和 I/O 设备的利用率，提高 I/O 操作速度，方便用户使用 I/O 设备。

设备管理包括缓冲区管理、设备分配、设备驱动等功能。

1. 缓冲区管理

CPU 运行高速性及 I/O 低速性之间的矛盾自计算机诞生时起便已存在了。而随着 CPU 速度迅速提高，使得此矛盾更为突出，严重降低了 CPU 的利用率。如果在 I/O 设备及 CPU 之间引入缓冲区，则可有效地缓和 CPU 及 I/O 设备之间速度不匹配的矛盾，提高 CPU 的利用率，进而提高系统吞吐量。因此，在现代计算机系统中，都无一例外地在内存中设置了缓冲区，而且可以通过增加缓冲区容量的方法改善系统的性能。

2. 设备分配

设备分配的基本任务是根据用户进程的 I/O 中断请求及系统的现有资源情况按照某种设备分配的策略，为用户分配其所需的设备。如果在 I/O 设备和 CPU 之间还存在着设备控制器及 I/O 通道，还必须为分配出去的设备分配相应的控制器及通道。

3. 设备驱动

设备驱动的作用是屏蔽底层硬件设备的差异性，向上层提供统一的操作及管理接口，以实现 CPU 及设备控制器之间的通信。设备驱动能够将 CPU 发出的 I/O 操作转换成设备控制器可以识别的 I/O 指令，也能够将设备控制器发来的中断请求反馈给 CPU 并给予迅速的响应和相应的处理。

1.4 操作系统的特点

1.4.1 并发性

并行性(Parallelism)和并发性(Concurrency)是既相似又有区别的两个概念，并行性是指两个或多个事件在同一时刻发生；并发性是指两个或多个事件在同一时间间隔内发生。串行(Sequence)，顾名思义，是指多个进程按顺序运行，进程串行、并发及并行的对比见图 1-7。

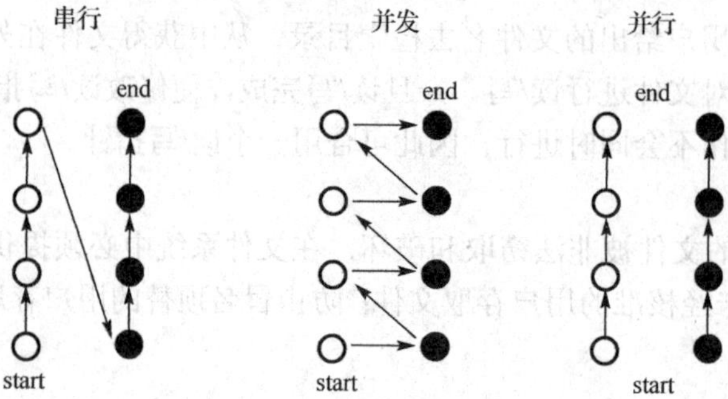

图 1-7　进程串行、并发和并行对比图

图 1-7 中的圆形代表不同的进程，箭头方向指示的是进程的运行顺序。根据多道程序设计思想，内存中同时存在多个进程，如果按串行方式运行，那么一个进程结束后，才会开始另一个进程的运行；如果按并发方式运行，那么每个进程不是一口气运行到底的，而是“走走停停”的，你运行一会儿，我运行一会儿，交替运行直至结束；如果按并行方式运行，那么两个完成相同任务的进程同时开始运行，同时结束(理想情况下)。

出现上述几种情况的根本原因在于计算机的 CPU 数量不足，在只有一个 CPU 的情况下，多个进程之间只能采取串行或并发的方式运行，如果有多个 CPU，则可以让多个进程并行作业。现代计算机虽然已经配备了多核 CPU，但内存中同时存在的进程数量远远大于 CPU 的核心数量，操作系统总是按并发方式运行进程的，因此并发性是操作系统的重要特征。

1.4.2　共享性

在操作系统环境下，所谓共享(Sharing)，是指系统中的资源可供内存中多个并发运行的进程共同使用，相应地，把这种资源的共同使用称为资源共享或资源复用。由于各种资源的属性不同，进程资源共享的方式也不同，目前实现资源共享的方式主要有如下两种。

1. 互斥共享方式

系统中的某些资源(如打印机)，虽然可以提供给多个进程(线程)使用，但为使(打印)结果不混淆，应规定在一段时间内只允许一个进程(线程)访问该资源。为此，系统中应建立一种机制，以保证对这类资源的互斥访问。例如，当一个进程 A 要访问某资源时，必须先提出请求。如果此时该资源空闲，系统便可将之分配给进程 A 使用。此后若再有其他进程也要访问该资源(A 未用完)，则必须等待。仅当进程 A 访问完并释放该资源后，才允许另一个进程对该资源进行访问。我们把这种资源共享的方式称为互斥共享，而把在一段时间内只允许一个进程(线程)访问的资源称为临界资源或独占资源。计算机系统中的大多物理设备，以及某些软件中所用的栈、变量及表格，都属于临界资源，它们要求被互斥地共享。为此，在系统中必须配置某种机制来保证多个进程(线程)互斥地使用临界资源。

2. 同时访问方式

系统中还有一类资源，允许在一段时间内由多个进程同时对它进行访问。这里所谓的“同时”，在单道批处理环境下往往是宏观上的，而在微观上，这些进程可能交替地对该资源进行访问。典型的可供多个进程"同时"访问的资源是磁盘，一些用重入码编写的文件也可以被"同时"共享。

并发及共享是操作系统的两个最基本的特征，它们又互为存在的条件。一方面，资源共享是以进程的并发运行为条件的，若系统不允许进程并发运行，自然不存在资源共享问题；另一方面，若系统不能对资源共享实施有效管理，协调好诸多进程对共享资源的访问，也必然影响到进程并发运行的程度，甚至使进程根本无法并发运行。

1.4.3 虚拟性

操作系统中的虚拟(Virtuality)是指通过某种技术把一个物理实体变为若干逻辑上的对应物。物理实体是实的，即实际存在的，而逻辑上的对应物是虚的，仅存在于用户的感觉之中。操作系统中使用了多种虚拟技术，如虚拟处理器、虚拟存储及虚拟外部设备(外设)等。

(1)虚拟处理器。为了让并发的进程分时共享一个物理处理器，操作系统为每个进程分配一个虚拟处理器，让每个进程都感觉自己独占处理器。

(2)虚拟存储器。一般是指将一种慢速大容量存储器虚拟成一种快速小容量存储器，最常见的是将磁盘空间虚拟成内存空间，达到扩充快速存储器的目的。内存可以作为缓存的虚拟存储器。

(3)虚拟外设。这是一种将一台物理设备虚拟成逻辑上的抽象设备的技术，如 SPOOLing 技术，它利用磁盘空间为中介，虚拟出多个低速的 I/O 设备，这样的设备被称为虚拟外设。

1.4.4 异步性

异步性(Asynchronism)又称随机性(Randomness)，即事件的发生无法预测。当操作系统接管计算机后，面对的都是随机事件，它无法预测鼠标何时被单击，键盘何时被敲击，以及进程要运行多长时间等，这些异步事件给操作系统的设计带来了极大的挑战。对于并发进程而言，如果对异步运行的进程不加以协调，会出现运行结果出错或不唯一等不良后果，因此操作系统提供多种同步工具协调异步进程的运行顺序以保证结果正确。

1.5 操作系统提供服务的方式

获取视频

1.5.1 接口

"接口"从字面上理解是两个事物之间的连接处，作用是将两个事物连接起来。该词在生活中其实很常见，如家用电源插座就是电器的电源插头和电源之间的接口。在计算机系统中，我们经常用到的 USB 设备有一端是凸出的方形 USB 接头，要与机器上凹形的 USB 插口相连接才可能正常工作，这个也是接口，因为这个接口连接了两个硬件，所以称为"硬件-硬件"接口，除此之外，计算机中的接口还有"硬件-软件"接口及"软件-软件"接口。

我们已经知道操作系统是一种系统软件，它与计算机硬件之间的接口就是"硬件-软件"接口。这层接口的实现分为两个层面：一是硬件提供操作的硬件指令集，如 CPU 指令集及 I/O

设备指令集等；二是由软件向硬件发出正确的指令，指挥硬件按照既定的方式工作。如果操作系统可以完全理解计算机的硬件指令集，并能够正确地发布指令及接收结果，那么用户就不必熟悉晦涩难懂的硬件指令，取而代之的是向操作系统发布相应的指令，然后操作系统将用户的指令转换成对硬件的对应指令，从而完成用户操作硬件的目的。因此，从另一个角度来看，操作系统扮演着中间人的角色，向用户提供一台友好的虚拟计算机，接受用户键盘或鼠标发布的任务，将这些任务转换成硬件指令后发送给硬件并响应硬件输出结果。这样的目的是降低用户操作计算机硬件的难度，操作系统设计得越友好，用户操作起来就越便捷。

用户和操作系统之间的接口称为“软件-软件”接口，操作系统提供两类这样的接口供用户使用：一类是操作接口，方便用户操作、使用计算机；另一类是程序接口，为用户编写程序提供各类底层服务。这两类接口也是操作系统向用户提供服务的两种方式。

1.5.2 操作接口

操作接口为用户提供操作计算机的手段，一般有指令行接口（Command Line Interface，CLI）和图形用户接口（Graphic User Interface， GUI）。

Ubuntu（一种 Linux 的发行版本）的 CLI 界面如图 1-8 所示。

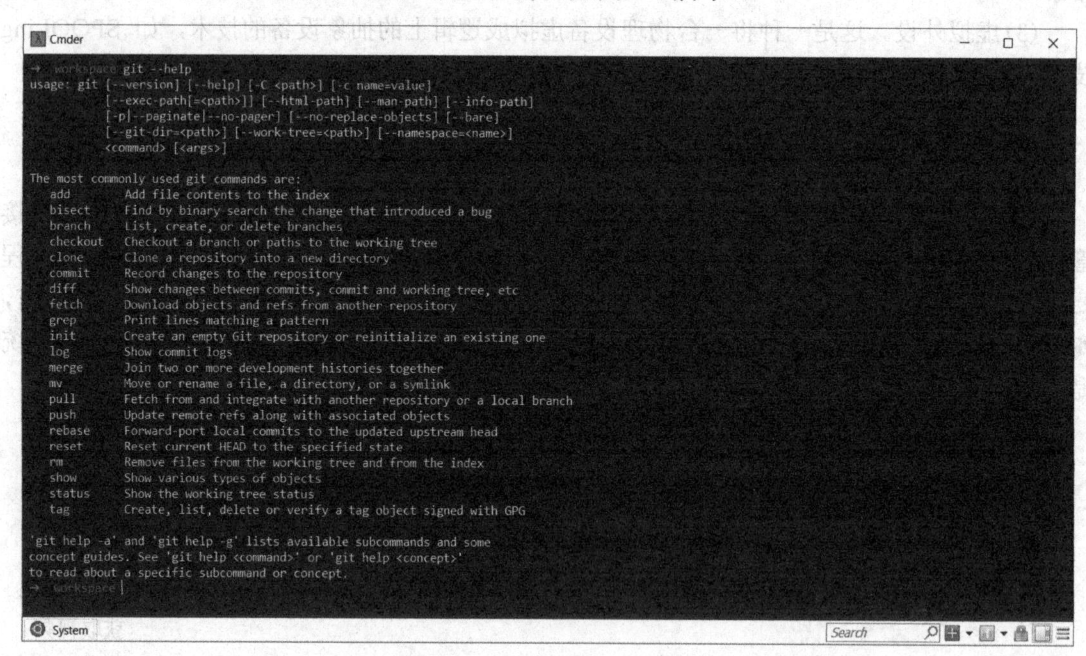

图 1-8　Ubuntu 的 CLI 界面

用户在终端窗口或控制台上输入一条指令后，系统便立即转入指令，对该指令加以解释并执行。在完成指定功能后，系统又回到终端或控制台上，等待用户输入下一条指令。这样，用户可通过先后输入不同指令的方式，来实现对作业的控制，直至任务完成。这种方式要求用户熟记各种指令的名字及参数，并严格按照规定的格式输入指令，既不方便，又费时间，于是简单易用的GUI方式被广泛采用。

GUI 采用了图形化的操作界面(如图 1-9 所示)，用非常容易识别的各种图标(Icon)将系统的各项功能、各种应用程序及文件直观、逼真地展示出来。用户可用鼠标或通过菜单和对话框来完成对应用程序及文件的操作。此时，用户已完全不必像使用指令那样去记住指令的名字及格式，从而把用户从烦琐且单调的操作中解脱出来。最早的图形用户接口于 1973 年应用在 Xerox Alto 计算机上(如图 1-10 所示)，1983 年的 Apple Lisa 计算机和 1984 年的 Apple Macintosh 引入了更多的图形化操作概念，如菜单及窗口控制等(如图 1-11 所示)，现在装机量最多的图形化操作系统非 Microsoft 的 Windows 莫属。

图 1-9　Ubuntu 的 GUI 界面

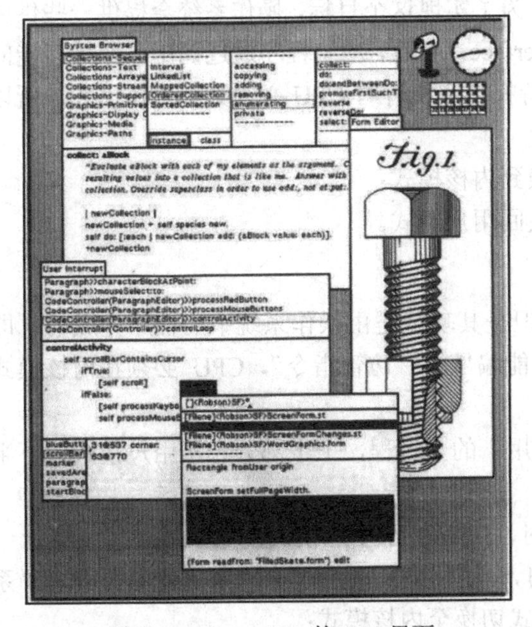

图 1-10　Xerox Alto 的 GUI 界面

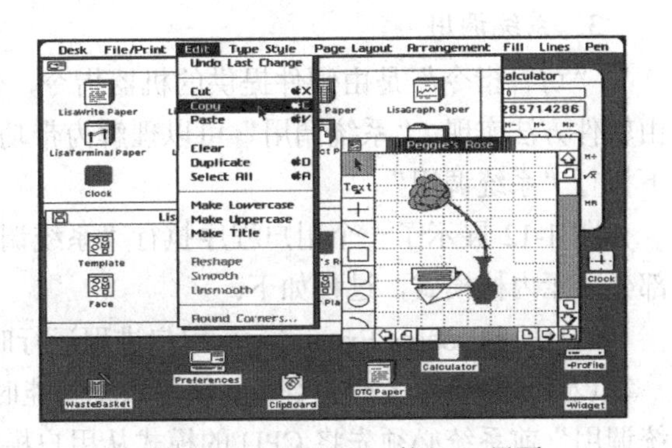

图 1-11　Apple Lisa 的 GUI 界面

1.5.3　程序接口

1. 概念

在操作系统的内核中都设置了用于实现各种系统功能的函数(某种可供重复使用的软件模块)，并将它们提供给应用程序调用，如向显示器发送字符串让它们显示在屏幕上。于是，用户编写程序时，不必考虑如何与显示器交互，只需要使用操作系统提供的功能函数就可以实现在显示器上输出字符串了。这些功能函数的调用接口统称为“程序接口”，因为这些系统功能是操作系统本身的一部分，为防止被恶意破坏，一般不允许用户直接访问，所以系统提供了一系列的“系统调用”(System Call)指令来让用户的应用程序调用这些系统功能函数，程序接口通常是由各种类型的“系统调用”组成。

2. 特权指令和CPU工作模式

特权指令是指有特殊权限的指令，这类指令的权限很大，如果使用不当，将导致整个系统崩溃，如清内存、置时钟、分配系统资源、修改虚拟内存的段表/页表及修改用户的访问权限等。如果所有的应用程序都能使用这类指令，那么对整个系统而言是非常危险的。为了保证系统安全，这类指令只能由操作系统或系统软件发出，不直接提供给用户程序使用。

为了能让CPU认识哪些指令是特权指令，操作系统人为地给CPU设置了工作模式。最简单的情形是有两种模式：用户模式(User Mode)和内核模式(Kernel Mode)。若CPU当前的工作状态为内核模式，则可以执行所有指令(特权和非特权指令)；若CPU处于用户模式，则只能执行非特权指令。换句话说，特权指令必须在CPU处于内核模式时才能被执行。

一般来说，应用程序不可以执行特权指令，因此它运行时，CPU工作在用户模式下，如果要执行特权指令，必须得到操作系统的许可。为了实现这个目标，操作系统会提供一些供应用程序使用的接口(Application Programming Interface，API)，应用程序通过调用这些接口能间接地执行操作系统才可以执行的指令。又因为特权指令执行时CPU必须处于内核模式，所以这个过程中包含两个模式切换：

(1)执行特权指令前，CPU从用户模式切换到内核模式；

(2)执行特权指令后，CPU从内核模式切换回用户模式。

3. 系统调用

“访管指令”是由硬件提供的机器指令，但是其功能是由操作系统中的程序完成的，即由软件方法实现。“系统调用”可以理解为带功能编号的“访管指令”，CPU必须在内核模式下执行“系统调用”。

图1-12展示了一个用户程序执行“系统调用”的全过程，上半部分表示用户模式，下半部分表示内核模式，过程如下：

(1)User Process Executing：用户进程运行时，CPU处于用户模式；

(2)Get System Call：当需要使用系统功能时，调用相应功能编号的“系统调用”，执行“系统调用”前系统必须先将CPU的模式从用户模式切换至内核模式；

(3) Execute System Call：在内核模式下执行“系统调用”；

(4) Return From System Call：“系统调用”执行完毕，CPU从内核模式切换回用户模式，继续运行用户程序的剩余代码。

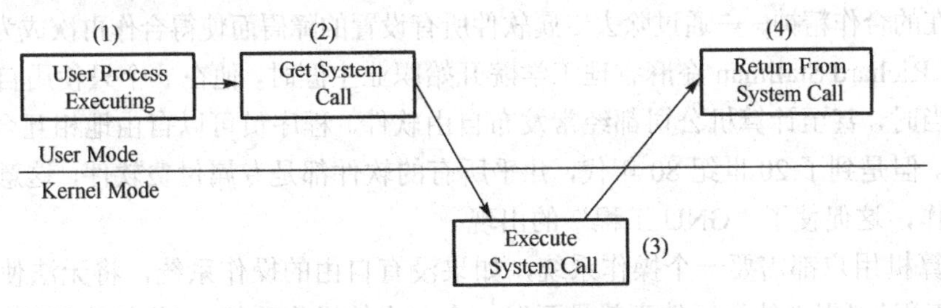

图1-12 系统调用

从第(2)步到第(3)步，有一个从用户模式向内核模式切换的过程，如果把图中央的横线比喻为一条马路，那么这个切换过程就像一个程序从路面陷入地下的过程，因此这个过程通常被形象地称为“陷入”，整个“系统调用”过程的机制被称为“陷阱机制”(Trap Mechanism)。

即使是一个简单的应用程序，也会包含非常多的“系统调用”动作。举个例子，假设一个程序的任务是从一个文件中读取数据并将数据复制到另一个文件中。这个程序首先需要获得这两个文件的文件名：输入文件及输出文件。获取文件名的方法有很多种，最直接的是向用户询问这两个文件的名称。这个过程将请求一系列“系统调用”。首先，在屏幕上显示提示信息。然后，读取由键盘输入的字符。一旦获得了这两个文件名，程序就要打开输入文件并创建输出文件。这些操作的每一步都需要请求另外的“系统调用”，因为打开文件和创建文件都涉及磁盘操作，这些动作不可以由应用程序直接完成。接下来，从输入文件中循环的读出数据(一个“系统调用”)，然后写入输出文件(另一个“系统调用”)。最后，在完成对整个文件的复制之后，程序就需要关闭这两个文件，向控制台中写一段消息(需要另外的“系统调用”)并最终正常结束(最后的“系统调用”)。

然而大多数用户从未见到过这些细节，取而代之的是一些简洁高效的API，比如，在C编程环境下打开文件用的是fopen()函数，读文件用的是fread()函数，写文件用的是fwrite()函数，关闭文件用的是fclose()函数。这些函数被包含在运行库当中，它降低了程序员使用“系统调用”的难度，为用户隐藏了操作系统接口的大部分细节。

1.6 GNU/Linux 历史

GNU系统是一套向上兼容UNIX的完全自由的操作系统，Richard Stallman在1983年9月做出了GNU工程的初始声明，接着又在1985年3月发表了更长的版本，叫GNU宣言，它被翻译成多种语言。叫“GNU”这个名字是因为它满足了几个要求：第一，它是“GNU's Not UNIX”的递归缩写；第二，它是一个真正的单词；第三，它说(或唱)起来有趣。

自由软件“free software”中的单词“free”关乎自由，而不是价格，所有人可以付费或不付费地得到GNU软件。无论如何，一旦得到了软件，用户便拥有了使用它的四项特定自由：

(1)有自由按照自己的意愿运行该软件；

(2)有自由复制软件并将其分享给朋友及同事；

(3)有自由通过对源代码的完全控制而改进程序；

(4)有自由发布改进的版本从而帮助社区建设。

开发 GNU 系统的工程叫“GNU 工程”。GNU 工程构思于 1983 年，意在找回早期计算机社区中广泛存在的合作精神——通过除去专属软件所有设置的障碍而使得合作再次成为可能。

1971 年 Richard Stallman 在麻省理工学院开始职业生涯时，他在一个只使用自由软件的小组中工作。当时，甚至计算机公司都经常发布自由软件。程序员可以自由地相互合作，他们也经常这样做。但是到了 20 世纪 80 年代，几乎所有的软件都是专属付费软件，这意味着禁止也阻止用户合作，这促使了“GNU 工程”的出现。

每个计算机用户都需要一个操作系统，如果没有自由的操作系统，将无法使用计算机，因此“GNU 工程”议程中的第一件事就是开发一个自由的操作系统。工作组选择了以兼容 UNIX 为基准，因为考虑到 UNIX 的整体设计历经考验并且可移植，同时兼容性也使得 UNIX 用户很容易从 UNIX 转移到 GNU 上。

一个类 UNIX 操作系统包括内核、编译器、编辑器、文本格式化软件、邮件软件、图形用户界面、应用库、游戏等。因此，编写一个完全的操作系统是一项巨大的工作。工作组从 1984 年 1 月开始设计开发，并在 1985 年 10 月创立自由软件基金会，为开发 GNU 募集资金。

到 1990 年，除了操作系统的内核，其他所有主要的系统组件均准备就绪。当时，Linus Torvalds 在 1991 年开发了一个类 UNIX 的内核 Linux，并在 1992 年将其变成自由软件。将 Linux 与几乎完成的 GNU 系统结合在一起就是一个完整的操作系统——GNU/Linux 系统。现在有几千万人在使用的 GNU/Linux 系统，通常是 GNU/Linux 的发行版。

然而，“GNU 工程”并不限于核心操作系统，它提供了一个完整的软件系列，以满足更多用户的任何要求，其中包括应用软件、图形用户界面(称为 GNOME)、游戏及其他娱乐应用。

1.7 Reading Materials

1.7.1 Overview

An operating system is software that manages the computer hardware, as well as providing an environment for application programs to run. Perhaps the most visible aspect of an operating system is the interface to the computer system it provides to the human user.

For a computer to do its job of executing programs, the programs must be in main memory. Main memory is the only large storage area that the processor can access directly. It is an array of bytes, ranging in size from millions to billions. Each byte in memory has its own address. The main memory is usually a volatile storage device that loses its contents when power is turned off or lost. Most computer systems provide secondary storage as an extension of main memory. Secondary storage provides a form of nonvolatile storage that is capable of holding large quantities of data permanently. The most common secondary-storage device is a magnetic disk, which provides storage of both programs and data.

The wide variety of storage systems in a computer system can be organized in a hierarchy according to speed and cost. The higher levels are expensive, but they are fast. As we move down the hierarchy, the cost per bit generally decreases, whereas the access time generally increases.

There are several different strategies for designing a computer system. Single-processor systems have only one processor, while multiprocessor systems contain two or more processors that share physical memory and peripheral devices. The most common multiprocessor design is symmetric multiprocessing (or SMP), where all processors are considered peers and run independently of one another. Clustered systems are a specialized form of multiprocessor systems and consist of multiple computer systems connected by a local-area network.

To best utilize the CPU, modern operating systems employ multiprogramming, which allows several jobs to be in memory at the same time, thus ensuring that the CPU always has a job to execute. Time-sharing systems are an extension of multiprogramming where in CPU scheduling algorithms rapidly switch between jobs, thus providing the illusion that each job is running concurrently.

The operating system must ensure correct operation of the computer system. To prevent user programs from interfering with the proper operation of the system, the hardware has two modes: user mode and kernel mode. Various instructions (such as I/O instructions and halt instructions) are privileged and can be executed only in kernel mode. The memory in which the operating system resides must also be protected from modification by the user. A timer prevents infinite loops. These facilities (dual mode, privileged instructions, memory protection, and timer interrupt) are basic building blocks used by operating systems to achieve correct operation.

A process (or job) is the fundamental unit of work in an operating system. Process management includes creating and deleting processes and providing mechanisms for processes to communicate and synchronize with each other.

An operating system manages memory by keeping track of what parts of memory are being used and by whom. The operating system is also responsible for dynamically allocating and freeing memory space. Storage space is also managed by the operating system; this includes providing file systems for representing files and directories and managing space on mass-storage devices.

Operating systems must also be concerned with protecting and securing the operating system and users. Protection measures control the access of processes or users to the resources made available by the computer system. Security measures are responsible for defending a computer system from external or internal attacks.

Several data structures that are fundamental to computer science are widely used in operating systems, including lists, stacks, queues, trees, hash functions, maps, and bitmaps.

Computing takes place in a variety of environments. Traditional computing involves desktop and laptop PCs, usually connected to a computer network. Mobile computing refers to computing on handheld smartphones and tablet computers, which offer several unique features. Distributed systems allow users to share resources on geographically dispersed hosts connected via a computer network. Services may be provided through either the client-server model or the peer-to-peer model. Virtualization involves abstracting a computer's hardware into several different execution environments. Cloud computing uses a distributed system to abstract services into a "cloud", where users may access the services from remote locations. Real-time operating systems are designed for embedded environments, such as consumer devices, automobiles, and robotics.

The free software movement has created thousands of open-source projects, including operating systems. Because of these projects, students are able to use source code as a learning tool. They can modify programs and test them, help find and fix bugs, and otherwise explore mature, full-featured operating systems, compilers, tools, user interfaces, and other types of programs.

GNU/Linux and BSD UNIX are open-source operating systems. The advantages of free software and open sourcing are likely to increase the number and quality of open-source projects, leading to an increase in the number of individuals and companies that use these projects.

1.7.2 Concurrency and Parallelism

Concurrency and parallelism are tow different things.

It may seem that there is no difference between concurrency and parallelism, but this is because you have not understood the essence of the question. Let's try to understand how they differ.

Concurrency（Fig.1）is the execution of more than one task is being processed in overlapping time periods. An important detail is that tasks are not necessarily performed at the same time(but it's possible). That is based on the notion of Interruptability — tasks can be divided into smaller and alternating subtasks. In this case, they can be executed simultaneously, but this is not necessary.

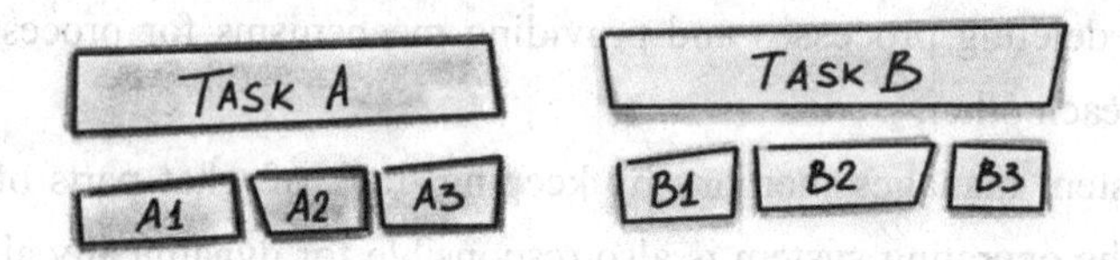

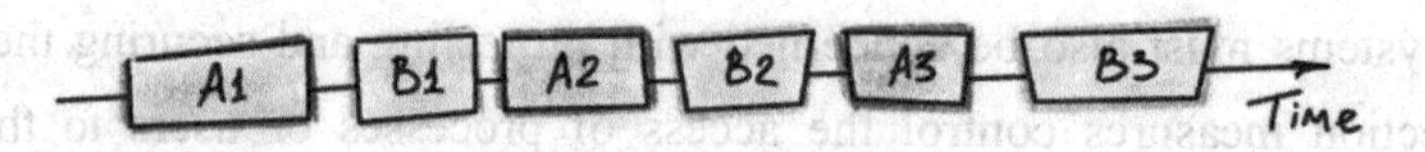

Fig.1 Concurrency

Subtasks are not connected with each other. Therefore, it does not matter which of them will end earlier, and which later. Thus, concurrency can be realized in many ways — using green threads or processes or asynchronous operations that work on one CPU or something else.

Let's draw an analogy: the secretary answers phone calls and sometimes checks for appointments. He needs to stop answering the phone to go to the desk and check the appointments, and then start answering and repeating the process before the end of the workday.

As you have noticed, concurrency is more connected with logistics. If it were not, then the secretary would wait until the time of the appointment and do the necessary things and then go to the ringing phone.

Parallelism (Fig.2) is literally the simultaneous execution of tasks. The very name implies that they are executed in parallel. Parallelism is one of the ways to implement concurrent execution highlighting abstraction of a thread or process. Also for parallelism to be true, there must be at least two computational resources.

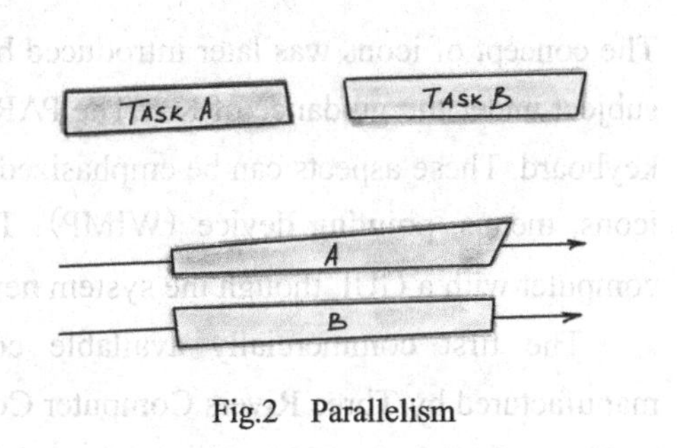

Fig.2 Parallelism

Back to the office: Now we have two secretaries. One keeps an eye on the phone, and the other makes appointments. The work is divided because now there are two secretaries in the office and the work is done in parallel.

Parallelism is a subclass of concurrency — before performing several concurrent tasks, you must first organize them correctly.

1.7.3 Graphic User Interface

The Graphical User Interface (GUI) is a form of user interface that allows users to interact with electronic devices through graphical icons and audio indicator such as primary notation, instead of text-based user interfaces, typed command labels or text navigation. GUIs were introduced in reaction to the perceived steep learning curve of Command Line Interfaces (CLIs), which require commands to be typed on a computer keyboard.

The actions in a GUI are usually performed through direct manipulation of the graphical elements. Beyond computers, GUIs are used in many handheld mobile devices such as MP3 players, portable media players, gaming devices, smartphones and smaller household, office and industrial controls. The term GUI tends not to be applied to other lower-display resolution types of interfaces, such as video games (where Head-Up Display (HUD) is preferred), or not including flat screens, like volumetric displays because the term is restricted to the scope of two-dimensional display screens able to describe generic information, in the tradition of the computer science research at the Xerox Palo Alto Research Center.

Early efforts

Ivan Sutherland developed Sketchpad in 1963, widely held as the first graphical computer-aided design program. It used a light pen to create and manipulate objects in engineering drawings in realtime with coordinated graphics. In the late 1960s, researchers at the Stanford Research Institute, led by Douglas Engelbart, developed the On-Line System (NLS), which used text-based hyperlinks manipulated with a then-new device: the mouse. (A 1968 demonstration of NLS became known as "The Mother of All Demos.") In the 1970s, Engelbart's ideas were further refined and extended to graphics by researchers at Xerox PARC and specifically Alan Kay, who went beyond text-based hyperlinks and used a GUI as the main interface for the Smalltalk programming language, which ran on the Xerox Alto computer, released in 1973. Most modern general-purpose GUIs are derived from this system.

The Xerox Star 8010 workstation introduced the first commercial GUI. The Xerox PARC user interface consisted of graphical elements such as windows, menus, radio buttons, and check boxes.

The concept of icons was later introduced by David Canfield Smith, who had written a thesis on the subject under the guidance of Kay.The PARC user interface employs a pointing device along with a keyboard. These aspects can be emphasized by using the alternative term and acronym for windows, icons, menus, pointing device（WIMP）. This effort culminated in the 1973 Xerox Alto, the first computer with a GUI, though the system never reached commercial production.

The first commercially available computer with a GUI was 1979 PERQ workstation, manufactured by Three Rivers Computer Corporation. Its design was heavily influenced by the work at Xerox PARC. In 1981, Xerox eventually commercialized the Alto in the form of a new and enhanced system — the Xerox 8010 Information System — more commonly known as the Xerox Star. These early systems spurred many other GUI efforts, including Lisp machines by Symbolics and other manufacturers, the Apple Lisa（which presented the concept of menu bar and window controls）in 1983, the Apple Macintosh 128K in 1984, and the Atari ST with Digital Research's GEM, and Commodore Amiga in 1985. Visi On was released in 1983 for the IBM PC compatible computers, but was never popular due to its high hardware demands. Nevertheless, it was a crucial influence on the contemporary development of Microsoft Windows.

Apple, Digital Research, IBM and Microsoft used many of Xerox's ideas to develop products, and IBM's Common User Access specifications formed the basis of the user interfaces used in Microsoft Windows, IBM OS/2 Presentation Manager, and the UNIX Motif toolkit and window manager. These ideas evolved to create the interface found in current versions of Microsoft Windows, and in various desktop environments for UNIX-like operating systems, such as macOS and Linux. Thus most current GUIs have largely common idioms.

Popularization

GUIs were a hot topic in the early 1980s. The Apple Lisa was released in 1983, and various windowing systems existed for DOS operating systems（including PC GEM and PC/GEOS）. Individual applications for many platforms presented their own GUI variants. Despite the GUIs advantages, many reviewers questioned the value of the entire concept, citing hardware limits, and problems in finding compatible software.

In 1984, Apple released a television commercial which introduced the Apple Macintosh during the telecast of Super Bowl XVIII by CBS, with allusions to George Orwell's noted novel Nineteen Eighty-Four. The goal of the commercial was to make people think about computers, identifying the user-friendly interface as a personal computer which departed from prior business-oriented systems, and becoming a signature representation of Apple products.

Windows 95, accompanied by an extensive marketing campaign, was a major success in the marketplace at launch and shortly became the most popular desktop operating system.

In 2007, with the iPhone and later in 2010 with the introduction of the iPad, Apple popularized the post-WIMP style of interaction for multi-touch screens, and those devices were considered to be milestones in the development of mobile devices.

The GUIs familiar to most people as of the mid-late 2010s are Microsoft Windows, macOS, and the X Window System interfaces for desktop and laptop computers, and Android, Apple's iOS,

Symbian, BlackBerry OS, Windows Phone/Windows 10 Mobile, Tizen, WebOS, and Firefox OS for handheld（smartphone）devices.

1.8 实验 1 Linux 安装及开发环境搭建

获取视频

1. 实验目的

操作系统理论是枯燥深奥的，仔细观察实验结果会帮助读者理解这些晦涩的理论知识，因此学习该课程的不二法门就是实验。好在安装一个 Linux 实验环境已经是非常简单的事情了，各种 Linux 发行版本和操作手册随处可见，这给初学者带来了极大的便利。作为全书的第一个实验，本实验的目的就是让初学者熟悉 Linux 环境，包括 Linux 安装、Linux 常用指令及 C 语言开发环境搭建。

2. 实验方法

(1) 选择一个 Linux 的发行版本。Linux 发展到现在，网络上有各种各样的发行版本，本书推荐读者选择国产的深度系统(Deepin)，其安装文件可在其官方网站上下载，不加特殊说明，本书中的所有代码默认是在该系统环境下运行的。

(2) Deepin 的安装：本书建议初学者使用虚拟机安装 Deepin，推荐的虚拟机软件是 VirtualBox，该软件是开源软件且同时适用于 Windows 及 macOS 系统。

(3) Linux 常用指令：在终端窗口中体验 CLI 方式，通过指令与操作系统进行交互，体验该方式和 GUI 方式的不同。

(4) C 语言开发环境搭建：本书的所有代码均用 C 语言编写，因此推荐安装 gcc 编译环境。

3. 实验内容

(1) Deepin 的下载及安装。

(2) Linux 常用指令：在应用列表中启动终端窗口(Terminal)，在 CLI 下了解 Linux 的目录结构及 Home 目录的作用，练习目录相关指令，包括但不限于以下指令：cd、pwd、mkdir、ls 及 gedit 等，指令的操作方法及说明请自行查阅资料学习。

(3) 搭建测试 gcc 环境，使用“sudo apt install gcc”指令安装 gcc 编译器及相关依赖环境。

(4) 在 home 目录中创建一个名为 practice 的子目录，用于存放实验代码。

(5) 进入 practice 目录中新建一个名为 helloworld.c 的文件，并将以下代码输入文件，保存后退出。

```
#include <stdio.h>
int main()
{
  printf("Hello, process\n");
  return 0;
}
```

(6) 使用 gcc 指令对 C 语言源文件进行编译和链接，最后运行并观察结果。

```
# gcc helloProcess.c -o helloProcess
#./helloProcess
Hello, process
```

第2章 进　　程

获取视频

2.1 程序和进程

程序是一种被动实体(Passive Entity)，它包含一组机器指令集，以文件的形式存放在外存上，我们通常称为“可执行文件”。不同操作系统的可执行文件格式不同，Windows 中的可执行文件通常以“exe”作为文件后缀名，Linux 中的可执行文件名没有明显特征，可通过查看文件属性来判断。早期的计算机一次只能运行一个程序，这个程序完全控制系统，并且能访问所有系统资源。相比之下，现代计算机系统允许加载多个程序到内存，以便并发运行。这种改进要求对各种程序提供更严格的控制及更好的划分。这些需求导致了进程概念的产生。

当程序文件中的指令集被加载入内存开始执行时，它的实体从外存转移到了内存中，此时的实体称为进程。进程是一种动态实体(Active Entity)，这个实体在内存中，除指令集之外，它还包含其他一些支撑指令执行的环境。进程是现代分时操作系统的工作单元，操作系统进程执行系统代码，而用户进程执行用户代码。通过 CPU 的多路复用，所有进程都可以并发运行。通过在进程之间切换 CPU，能使计算机更高效。

举个例子，下面的代码是用 C 语言编写的，在第 1 章实验 1 中我们练习了如何编写 C 语言源代码及使用 gcc 对其进行编译、链接，最终生成一个可执行文件，这个保存在磁盘上的可执行文件就是“程序”，当你使用“./”指令要求 Linux 执行它时，它就成了“进程”。

```
 1 int global=100;
 2
 3 void f(int x, int y){
 4   int* p = malloc(100);
 5   return;
 6 }
 7
 8 void g(int a){
 9   f(a, a+1);
10   return;
11 }
12
13 int main()
14 {
15   static int i=10;
16   g(i);
17   return 0;
18 }
```

需要注意的是，程序及进程不是相互独立的，而是有关联的，进程实体在内存中，程序文件中的一系列指令当然也包含在进程实体中，只不过除了这些，进程还需要额外的数据来支撑它的运行，其中最关键的一个就是程序计数器(Program Counter)，简称 PC。PC 是指一条指令的内存地址，存放在 CPU 的 PC 寄存器中，当 CPU 从该地址中取出指令之后，会自动将 PC 寄存器的值加 1，即表示下一条要执行指令的地址，如此重复，直至程序主函数返回。

不难看出，一个程序要运行，只有代码是不够的，支撑进程运行的内存环境可分成四部分（如图 2-1 所示）：代码（text）、数据（data）、堆（heap）及栈（stack），我们继续从上述代码出发，详细地解释这几部分的功能及作用。第 1 部分是代码部分，其作用不言而喻，就是存放可执行的指令序列，这部分在内存中一般是只读属性；第 2 部分是数据部分，用于存放全局变量及静态变量，第 1 行代码的 global 变量和第 15 行代码的 i 变量的空间就存放在数据部分；第 3 部分是堆部分，用于程序运行时的动态内存分配，代码的第 4 行用 malloc 申请的 100 字节的内存空间就来自于堆；第 4 部分是栈部分，这是一个先进后出的数据结构，按照指令序列的执行顺序将定义的局部变量（第 3 行代码的 x 和 y 变量，第 8 行代码的 a 变量）及被调用函数的返回地址压入（Push）栈，当函数返回时再按相反的顺序将栈顶数据弹出（Pop）。

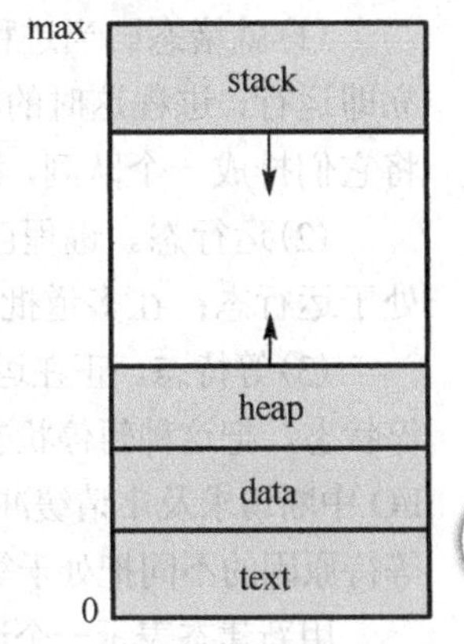

图 2-1　进程的内存结构

虽然两个进程可以与同一程序相关联，但是也可以被当成两个单独的执行序列。例如，多个用户可以运行电子邮件的不同副本，同一用户可以调用 Web 浏览器程序的多个副本。每个进程都是单独的进程，虽然文本段相同，但是数据段、堆及栈不同。一个进程在运行时也经常会生成许多进程。

最后我们给进程下一个定义：进程是一个程序的一次运行过程，它能够在某个数据集上完成具体的功能，它的运行过程可以是并发的；进程是资源分配、保护及调度的基本单位。

2.2　进程的状态及转换

进程的并发性是指多个进程同时驻留在内存中，在一段时间内，宏观上多个进程是同时运行的，但在单处理系统中，每个时刻只有一个进程可以占用处理器，因此微观上这些进程是分时交替运行的。如图 2-2 所示。

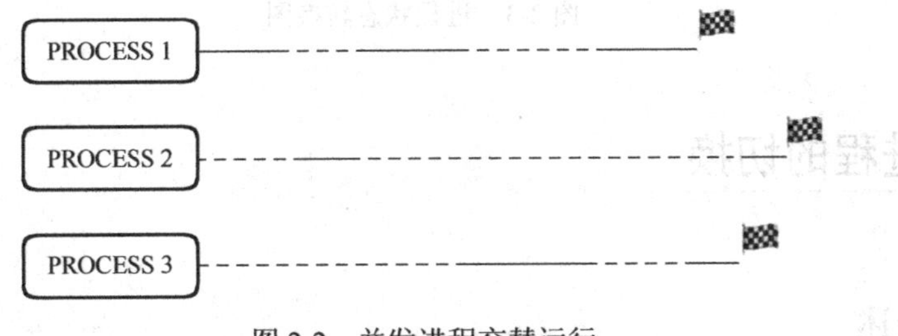

图 2-2　并发进程交替运行

在图 2-2 中，三个并发运行的进程中的实线表示进程占用 CPU 运行指令，虚线表示进程暂离开了 CPU，不难看出，在同一时刻只有一个进程占用 CPU（假设只有一个 CPU），进程的运行并不是一气呵成的，而是“走走停停”的，直至终点。

进程只有占用了 CPU 才可以执行它的功能指令，当一个进程占用了 CPU 后，有两种情况会导致进程失去 CPU 的使用权：进程主动离开（Yield）或进程被迫离开（Preempt）。

进程主动离开：也称主动放弃 CPU。发生这种情况的原因可能是进程运行结束，不再需要 CPU；也可能是进程执行了 I/O 操作，要等待 I/O 操作结束才能继续运行，此时进程会主动让出 CPU 给其他进程使用。

进程被迫离开：也称被强行剥夺 CPU。一般发生在进程消耗完自己的时间后，操作系统强迫进程交出 CPU 的使用权，并将其调度出去。

进程运行时的间断性决定了进程可能具有多种状态。事实上，运行中的进程可能具有以下三种基本状态。

(1) 就绪态。当进程已被分配到除 CPU 以外的所有必要资源后，只要再获得 CPU，便可立即运行，进程这时的状态称为就绪态。在一个系统中，处于就绪态的进程可能有多个，通常将它们排成一个队列，称为就绪队列。

(2) 运行态。进程已获得 CPU，其程序正在运行。在单道批处理系统中，只能有一个进程处于运行态；在多道批处理系统中，则有多个进程处于运行态。

(3) 等待态。正在运行的进程由于发生某事件而暂时无法继续运行时，便放弃 CPU 而处于暂停状态，把这种暂停状态称为等待态，有时也称为阻塞态或封锁态。致使进程阻塞的典型事件有 I/O 中断请求及申请缓冲空间等。通常将这种处于等待态的进程也排成一个队列，有的系统则根据等待原因的不同把处于等待态的进程排成多个队列。

用新建态表示一个进程的创建，用终止态表示进程运行完毕，可以得到经典的进程状态转换图，如图 2-3 所示。

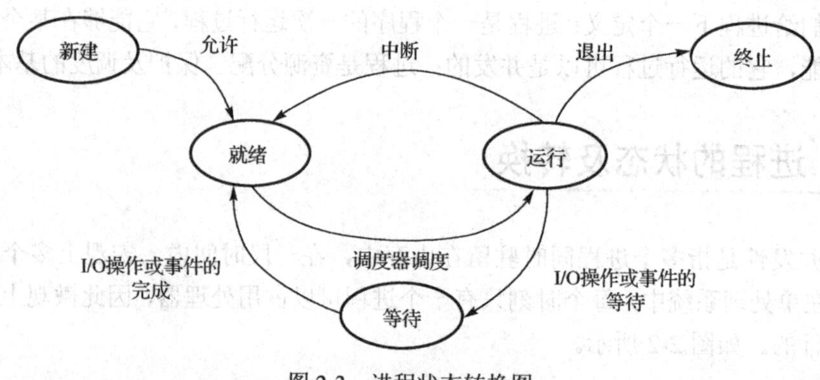

图 2-3　进程状态转换图

2.3　进程的切换

获取视频

2.3.1　概述

并发进程运行过程中，一个进程在运行时可能会被另一个进程替换占用 CPU，这个过程称为“进程切换”。现在提出两个问题：是什么触发了进程切换？进程切换时要做些什么？我们在 2.2 节中讨论了进程运行时离开 CPU 的原因，当发生所述的两种情况时会触发中断机制，在中断处理程序中完成进程切换。本节的内容就从中断机制开始，一步一步地深入介绍进程切换的完整过程。

2.3.2 中断机制

1. 中断

中断是计算机发展中的一种重要技术。最初它是为解决对I/O接口的控制采用程序查询所带来的CPU低效问题而产生的。简而言之，中断是指程序运行过程中，当发生某个事件时，中止CPU上现有程序的运行，运行该事件的中断处理程序，运行完毕后，中断处理程序返回原程序中断点继续运行，如图2-4所示。

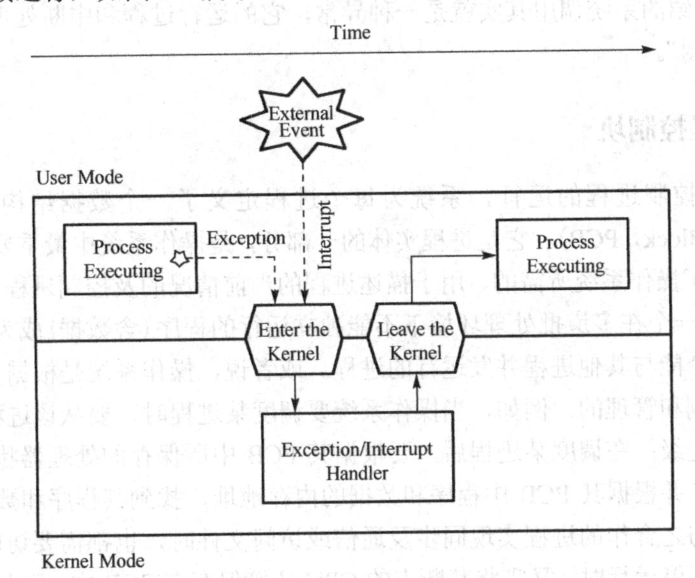

图2-4 中断处理流程图

在图2-4中，时间线(Time)从左至右推进，在一个进程运行过程中遇到一个事件(External Event)时，触发中断，转而运行中断处理程序(Interrupt Handler)，运行完毕后返回进程，继续运行。

2. 中断源(Interrupt Sources)

引起中断的原因或者能够发出中断请求信号的来源统称为中断源。中断源有很多种，分类方法也不尽相同，本书按中断源来自CPU外部或内部的原则将中断源分为外中断(External Interrupt)及内中断(Internal Interrupt)两类。

外中断是来自CPU外部的硬件中断信号，如时钟中断、键盘中断及外围设备中断等，这些均为异步中断。在不特殊说明的情况下，“Interrupt”一词就是指外中断，请注意图2-4中的标注。

内中断来自CPU内部，通常指指令执行过程中发生的中断，属于同步中断，如奇偶校验错误、程序指令非法操作、地址越界、断点或除数为零及系统调用引起的中断等。内中断一般称为“异常”(Exception)，请注意图2-4中的标注。

3. 中断处理程序

中断处理程序也称中断服务例程(Interrupt Service Routine，ISR)，通常由软件实现，它的处理内容和中断源有关，如处理除数为零的异常，就会发出计算警告并要求系统强制停止程序的继续运行。

中断源和中断处理程序共同构成了中断系统。

2.3.3 模式切换

因为中断处理程序中的指令大多和硬件相关，所以中断处理程序运行时要求CPU必须处于内核模式下，又因为在进入中断处理程序之前，进程处于用户模式下，所以当发生中断或异常时，会出现一次CPU的模式切换。在图2-4中，当发生Interrupt或Exception时，会首先进入内核模式(Enter the Kernel)，中断处理程序运行完毕后，会通过一个硬件指令将CPU的模式切换回用户模式(Leave the Kernel)。

1.5.3节中介绍的系统调用其实就是一种异常，它的运行过程和中断处理方式是完全相同的，即“陷阱机制”。

2.3.4 进程控制块

为了描述和控制进程的运行，系统为每个进程定义了一个数据结构——进程控制块(Process Control Block，PCB)，它是进程实体的一部分，是操作系统中最重要的记录型数据结构。PCB中记录了操作系统所需的、用于描述进程的当前情况的及控制进程运行的全部信息。PCB的作用是使一个在多道批处理环境下不能独立运行的程序(含数据)成为一个能独立运行的基本单位，一个能与其他进程并发运行的进程。或者说，操作系统是根据PCB来对并发运行的进程进行控制和管理的。例如，当操作系统要调度某进程时，要从该进程的PCB中查出其现行状态及优先级；在调度某进程后，要根据其PCB中所保存的处理器状态信息，设置该进程运行的情况，并根据其PCB中程序和数据的内存地址，找到其程序和数据；进程在运行过程中，需要和与之合作的进程实现同步及通信或访问文件时，也都需要访问PCB；当进程由于某种原因而暂停运行时，又需将其断点的CPU环境保存在PCB中。可见，在进程的整个生命周期中，操作系统总是通过PCB对进程进行控制的，所以PCB是进程存在的唯一标志。

当操作系统创建一个新进程时，就为它建立了一个PCB；进程结束时操作系统会回收其PCB，进程也随之消亡。PCB可以被操作系统中的多个模块读或修改，如调度程序、资源分配程序、中断处理程序及监督和分析程序。因为PCB经常被访问，尤其是被运行频率很高的进程访问，所以PCB应常驻内存。操作系统将所有的PCB组织成若干链表(或队列)，存放在操作系统专门开辟的PCB区内。

一个进程的上下文(Context)是指与该进程相关的所有信息，它们全部保存在PCB中，主要信息如下。

(1)进程状态(Process State)：包括新建、就绪、运行、等待及终止等。

(2)程序计数器(Program Counter)：程序计数器表示进程将要执行的下一条指令的地址。

(3)CPU寄存器(CPU Register)：根据计算机体系结构的不同，CPU寄存器的类型和数量也会有所不同。它们包括累加器、索引寄存器、堆栈指针寄存器、通用寄存器及其他条件码信息寄存器。在发生中断时，这些状态信息与程序计数器一起保存，以便进程以后能正确地继续运行。

(4)CPU调度信息(CPU-Scheduling Information)：这类信息包括进程优先级及调度队列的指针等调度参数。

(5)内存管理信息(Memory-Management Information)：这类信息包括基地址、界限寄存器的值、页表或段表等。

(6) 记账信息（Accounting Information）：这类信息包括 CPU 时间、实际使用时间、时间期限、记账数据及作业或进程数量等。

(7) I/O 状态信息（I/O Status Information）：这类信息包括分配给进程的 I/O 设备列表及打开的文件列表等。

简而言之，PCB 简单地作为这些信息的仓库，这些信息随着进程的不同而不同。通常，通过状态保存（State Save）来保存 CPU 当前状态；之后，通过状态恢复（State Restore）使进程重新开始运行。

2.3.5 进程切换

切换 CPU 到另一个进程需要保存当前进程状态及恢复另一个进程的状态，这个任务称为进程切换（Process Switch），也叫上下文切换（Context Switch）。当进行进程切换时，旧进程状态会保存在其 PCB 中，然后加载经调度而要运行的新进程。

进程切换的详细步骤见图 2-5。

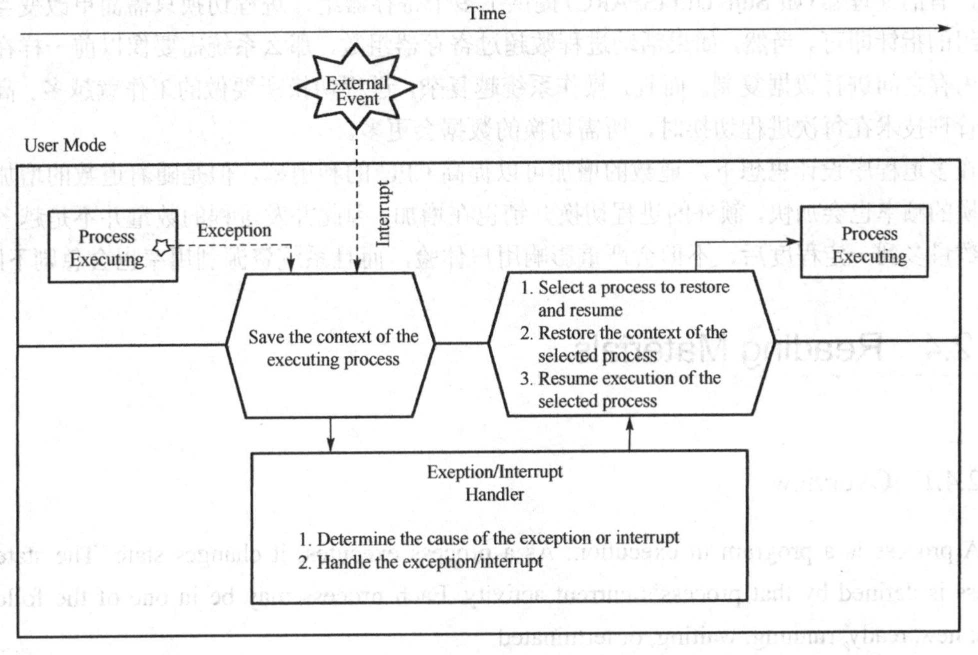

图 2-5 进程切换的详细步骤

当中断发生时：

(1) CPU 会从用户模式切换到内核模式，同时将当前正在运行进程的上下文信息保存在内存中；

(2) 在内核模式下，中断处理程序依据中断源的不同完成不同的工作，在处理进程切换时需要修改被中断进程的控制信息（如进程状态等），将被中断进程加入相应的状态队列中；

(3) 选择一个新的进程准备恢复，将它的上下文信息从内存中恢复，将 CPU 切换回用户模式，最后将 CPU 的使用权交给新进程，完成进程切换。

进程切换的时间是纯粹的开销，如图 2-6 所示。

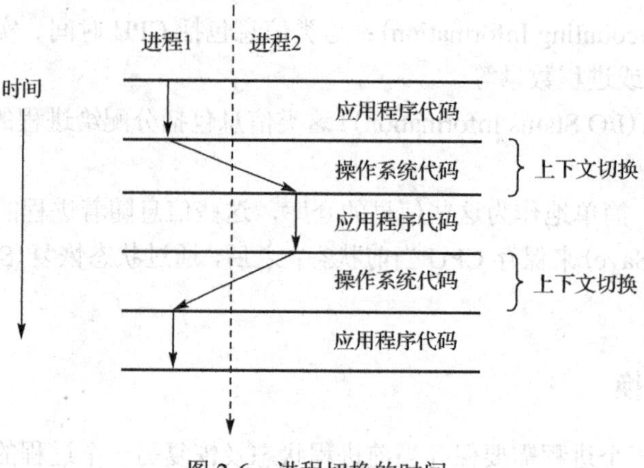

图 2-6　进程切换的时间

在切换时，系统没有做任何有用工作。进程切换的速度因计算机的不同而不同，它依赖于内存速度、必须复制的寄存器数量及是否有特殊指令。进程切换的时间与硬件支持密切相关。例如，有的处理器(如 Sun UltraSPARC)提供了多个寄存器组，进程切换只需简单改变当前寄存器组的指针即可。当然，如果活动进程数超过寄存器组数，那么系统需要像以前一样在寄存器与内存之间进行数据复制。而且，操作系统越复杂，进程切换所要做的工作就越多。高级的内存管理技术在每次进程切换时，所需切换的数据会更多。

在多道程序设计思想下，道数的增加可以提高 CPU 的利用率，但是随着道数的增加，进程切换的频率也会加快，额外的进程切换开销也在增加，因此并发进程的数量并不是越多越好的，数量多到一定程度后，不但会严重影响用户体验，而且系统资源利用率也会急剧下降。

2.4　Reading Materials

2.4.1　Overview

A process is a program in execution. As a process executes, it changes state. The state of a process is defined by that process's current activity. Each process may be in one of the following states: new, ready, running, waiting, or terminated.

Each process is represented in the operating system by its own Process Control Block (PCB).

A process, when it is not executing, is placed in some waiting queue. There are two major classes of queues in an operating system: I/O request queues and the ready queue. The ready queue contains all the processes that are ready to execute and are waiting for the CPU. Each process is represented by a PCB.

The operating system must select processes from various scheduling queues. Long-term (job) scheduling is the selection of processes that will be allowed to contend for the CPU. Normally, long-term scheduling is heavily influenced by resource-allocation considerations, especially memory management. Short-term (CPU) scheduling is the selection of one process from the ready queue.

Operating systems must provide a mechanism for parent processes to create new child processes. The parent may wait for its children to terminate before proceeding, or the parent and children may execute concurrently. There are several reasons for allowing concurrent execution: information sharing, computation speedup, modularity, and convenience.

The processes executing in the operating system may be either independent processes or cooperating processes. Cooperating processes require an interprocess communication mechanism to communicate with each other. Principally, communication is achieved through two schemes: shared memory and message passing. The shared-memory method requires communicating processes to share some variables. The processes are expected to exchange information through the use of these shared variables. In a shared-memory system, the responsibility for providing communication rests with the application programmers; the operating system needs to provide only the shared memory. The message-passing method allows the processes to exchange messages. The responsibility for providing communication may rest with the operating system itself. These two schemes are not mutually exclusive and can be used simultaneously within a single operating system.

Communication in client-server systems may use (1) sockets, (2) Remote Procedure Calls (RPCs), or (3) pipes. A socket is defined as an endpoint for communication. A connection between a pair of applications consists of a pair of sockets, one at each end of the communication channel. RPCs are another form of distributed communication. An RPC occurs when a process (or thread) calls a procedure on a remote application. Pipes provide a relatively simple ways for processes to communicate with one another. Ordinary pipes allow communication between parent and child processes, while named pipes permit unrelated processes to communicate.

2.4.2 Inter Process Communication

A process can be of two types: independent process and co-operating process.

An independent process is not affected by the execution of other processes while a co-operating process can be affected by other executing processes. Though one can think that those processes, which are running independently, will execute very efficiently, in reality, there are many situations when co-operative nature can be utilised for increasing computational speed, convenience and modularity. Inter Process Communication (IPC) is a mechanism which allows processes to communicate with each other and synchronize their actions. The communication between these processes can be seen as a method of co-operation between them. Processes can communicate with each other through both: Shared Memory and Message passing.The Fig.1 below shows a basic structure of communication between processes via the shared memory method and via the message passing method.

An operating system can implement both method of communication. First, we will discuss the shared memory methods of communication and then message passing. Communication between processes using shared memory requires processes to share some variable and it completely depends on how programmer will implement it. One way of communication using shared memory can be imagined like this: Suppose process1 and process2 are executing simultaneously and they share some resources or use some information from another process. Process1 generate information about certain computations or resources being used and keeps it as a record in shared memory. When process2 needs to use the shared information, it will check in the record stored in shared memory and take note of the information generated by process1 and act accordingly. Processes can use shared memory for extracting information as a record from another process as well as for delivering any specific information to other processes.

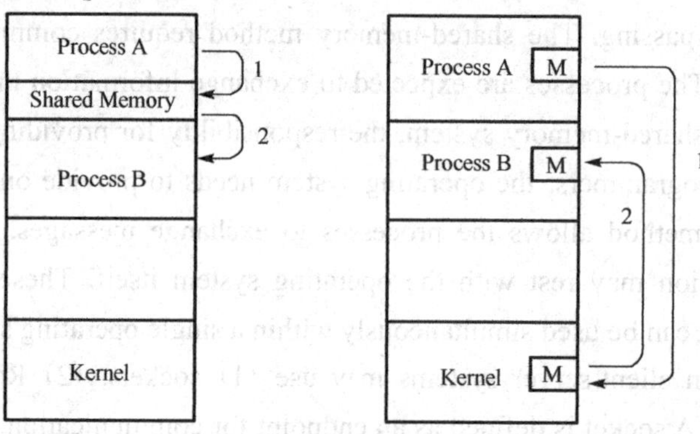

Fig.1 A basic structure of communication between processes via the shared memory method and via the message passing method

2.4.3 Process Control Block

A Process Control Block (PCB) is a data structure used by computer operating systems to store all the information about a process. It is also known as a process descriptor. When a process is created (initialized or installed), the operating system creates a corresponding process control block.

Information in a process control block is updated during the transition of process states. When the process terminates, its PCB is returned to the pool from which new PCBs are drawn.

Each process has a single PCB.

Role

The role of the PCBs is central in process management: they are accessed and/or modified by most utilities, particularly those involved with scheduling and resource management.

Structure

In multitasking operating systems, the PCB stores data needed for correct and efficient process management. Though the details of these structures are system-dependent, common elements fall in three main categories:

- Process identification
- Process state
- Process control

Status tables exist for each relevant entity, like describing memory, I/O devices, files and processes.

Memory tables, for example, contain information about the allocation of main and secondary (virtual) memory for each process, authorization attributes for accessing memory areas shared among different processes, etc. I/O tables may have entries stating the availability of a device or its assignment to a process, the status of I/O operations, the location of memory buffers used for them, etc.

Process identification data include a unique identifier for the process (almost invariably an integer) and, in a multiuser-multitasking system, data such as the identifier of the parent process, user identifier, user group identifier, etc. The process id is particularly relevant since it is often used to cross-reference the tables defined above, e.g. showing which process is using which I/O devices, or memory areas.

Process state data define the status of a process when it is suspended, allowing the OS to restart it later. This always includes the content of general-purpose CPU registers, the CPU process status word, stack and frame pointers, etc. During context switch, the running process is stopped and another process runs. The kernel must stop the execution of the running process, copy out the values in hardware registers to its PCB, and update the hardware registers with the values from the PCB of the new process.

Process control information is used by the OS to manage the process itself. This includes:

- Process scheduling state-The state of the process in terms of "ready" "suspended", etc., and other scheduling information as well, such as priority value, the amount of time elapsed since the process gained control of the CPU or since it was suspended. Also, in case of a suspended process, event identification data must be recorded for the event the process is waiting for.
- Process structuring information-the process's children id's, or the id's of other processes related to the current one in some functional way, which may be represented as a queue, a ring or other data structures.
- Inter-process communication information-flags, signals and messages associated with the communication among independent processes.
- Process Privileges-allowed/disallowed access to system resources.
- Process State-new, ready, running, waiting, dead.
- Process Number (PID)-unique identification number for each process (also known as Process ID).
- Program Counter (PC)-A pointer to the address of the next instruction to be executed for this process.
- CPU Registers-Register set where process needs to be stored for execution for running state.
- CPU Scheduling Information-information scheduling CPU time.
- Memory Management Information-page table, memory limits, segment table.
- Accounting Information-amount of CPU used for process execution, time limits, execution ID, etc.
- I/O Status Information-list of I/O devices allocated to the process.

Location

PCB must be kept in an area of memory protected from normal process access. In some operating systems the PCB is placed at the beginning of the kernel stack of the process.

2.5 实验2 进程的创建

获取视频

1. 实验目的

理解创建子进程函数fork()的用法，通过观察代码及运行结果理解父子进程的基本特征，掌握创建子进程、区分父子进程、获得进程PID及简单的父子进程同步等方法。

2. 实验方法

本次实验属于验证型实验，按照实验内容完成所有实验步骤，并记录实验结果即可，遇到不懂的问题或在某一步骤上“卡壳”，先尝试在搜索引擎上寻找解决方法，或严格跟着演示视频一步一步地进行。

3. 实验内容

(1)使用gedit编辑器新建一个helloProcess.c源文件，并输入下面的示例代码。

```
1 #include <stdio.h>
2 #include <sys/types.h>
3 #include <unistd.h>
4 int main()
5 {
6     //pid_t是数据类型，实际上是整型，通过typedef重新定义了一个名字，用于存储进程id
7     pid_t pid,cid;
8     //getpid()函数返回当前进程的id
9     printf("Before fork Process id :%d\n", getpid());
10    /*
11    fork()函数用于创建一个新的进程，该进程为当前进程的子进程，创建的方法是将当前进程的内存内容完整
  复制一份到内存的另一个区域，两个进程为父子关系，它们会同时(并发)执行fork()语句后面的所有语句。
12    fork()函数的返回值：
13    如果成功创建子进程，父子进程fork()函数会返回不同的值，对于父进程，它的返回值是子进程的进
  程id，对于子进程，它的返回值是0。
14    如果创建失败，返回值为-1。
15    */
16    cid = fork();
17    printf("After fork, Process id :%d\n", getpid());
18    return 0;
19 }
```

保存并退出gedit，使用gcc指令对源文件进行编译，然后运行，观察结果并解释原因。

(2)练习ps指令。该指令可以列出系统中当前运行的进程状态，在代码的17行和18行之间加入下面两行语句，目的是让父子进程暂停下来，否则无法观察它们运行的状态。

```
int i;
scanf("%d",&i);
```

重新编译、运行程序，开启一个新的终端窗口，输入下面的指令并观察结果。

```
ps -al
```

(3) 通过判断 fork () 函数的返回值让父子进程执行不同的语句。

```
 1 #include <stdio.h>
 2 #include <sys/types.h>
 3 #include <unistd.h>
 4 int main()
 5 {
 6    pid_t cid;
 7    printf("Before fork process id :%d\n", getpid());
 8    cid = fork();
 9    if(cid == 0){ //该分支是子进程执行的代码
10       printf("Child process id (my parent pid is %d):%d\n", getppid(), getpid());
11       for(int i=0; i<3 ; i++)
12          printf("hello\n");
13     }else{ //该分支是父进程执行的代码
14       printf("Parent process id :%d\n", getpid());
15       for(int i=0; i<3 ; i++)
16          printf("world\n");
17 }
18    return 0;
19 }
```

重新编译、运行程序，观察结果，重点观察父子进程是否判断正确(通过比较进程 id)。父子进程其实是并发运行的，但实验结果看起来像是顺序运行的，多运行几遍，看看有无变化，如果没有变化，试着将两个循环的次数调整得再高一些，比如调整为 30、300，然后再观察运行结果并解释原因。

图 2-7 解释了 fork () 函数的工作流程，请参照代码仔细理解。

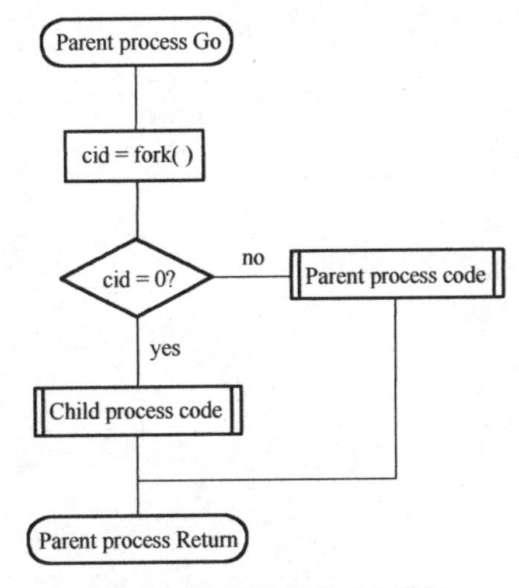

图 2-7 fork () 函数的工作流程

(4) 验证父子进程的内存空间是相互独立的。在终端窗口中进入主目录，使用 gedit 指令新建一个文件 helloProcess2.c，输入以下代码，然后编译、运行，观察运行结果并解释原因。

```
1 #include <stdio.h>
2 #include <sys/types.h>
3 #include <unistd.h>
4 int main()
```

```
 5 {
 6     pid_t cid;
 7     int x = 100;
 8     cid = fork();
 9     if(cid == 0){ //该分支是子进程执行的代码
10         x++;
11         printf("In child: x=%d\n",x);
12     }else{ //该分支是父进程执行的代码
13         x++;
14         printf("In parent: x=%d\n",x);
15     }
16     return 0;
17 }
```

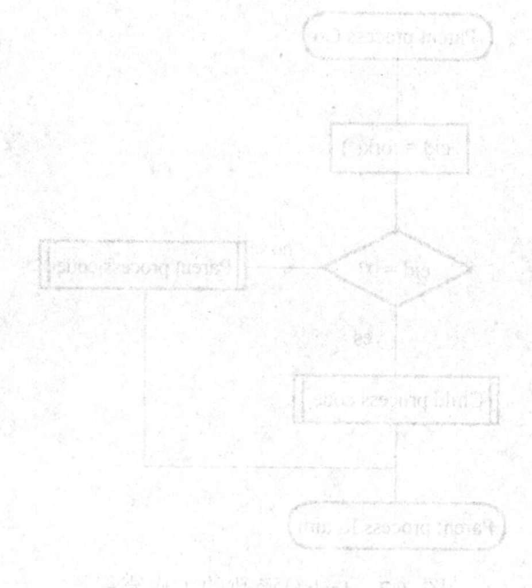

第3章 线 程

获取视频

3.1 动机和特点

并发进程是基于多道程序设计思想的，可以应用在一台独立的计算机上，也可以应用在如图3-1所示的客户端-服务器模型上，服务器(Server)运行的服务进程是为所有客户端(Client)服务的，服务器不断地监听客户端的请求，每当一个客户端向服务器发出请求时，服务器就会创建一个进程为之服务，根据第2章所学，进程的内存和其他占用资源是相互独立、不共享的，这个请求响应过程见图3-2。不过，创建进程很耗费时间和资源，一个Web服务器要接收有关网页、图像及声音等的客户请求，一个繁忙的Web服务器可能有多个(数百、数千个)客户并发访问它。如果新进程与原进程执行同样的任务，那么为什么要多承担这些开销？有没有可能服务器只创建一个进程来响应不同客户端的请求？

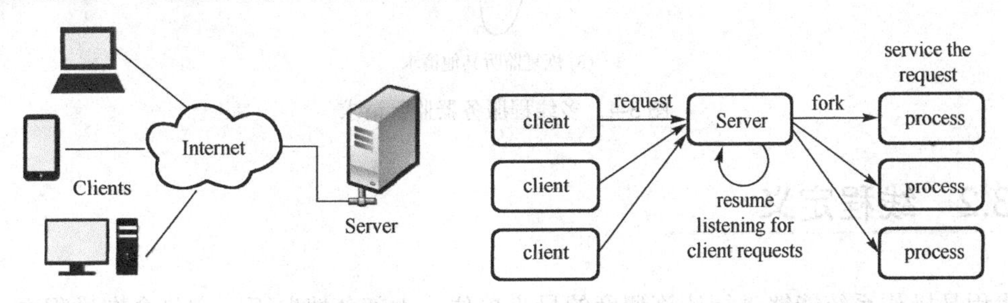

图3-1 客户端-服务器模型　　图3-2 请求响应过程

在20世纪60年代，人们提出进程的概念后，操作系统一直是以进程作为拥有资源和独立运行的基本单位。直到20世纪80年代中期，人们又提出了比进程更小且能独立运行的基本单位——线程(Thread)，试图用它来提高操作系统内程序并发运行的程度，从而进一步提高操作系统的吞吐量。特别是在进入20世纪90年代后，多处理器系统得到迅速发展，线程也能更好地提高程序的并发程度，充分地发挥多处理器的优越性。

举一个例子，如果有一段程序代码中有三个函数：main()、foo()及bar()，当该程序运行后，进程在内存中的实体如图3-3(a)所示。

PC表示程序计数器，指示当前执行指令的地址，从代码段来看，main()、foo()和bar()函数是按顺序执行的，如果使用fork()创建子进程，那就等同于将整个进程空间复制一份。现在我们做一些猜想，如果该进程中的三个函数可以并发执行，也就是说，一个进程中有三个执行流(函数)，按道理每个执行流都需要有一个PC，我们满足这个需求，为每个执行流分配一个PC，分别是PC1、PC2和PC3，当然我们只有一个CPU，这些执行流是并发执行的，也就是交替执行的[如图3-3(b)所示]。

这么做的好处在于，所有的执行流也具备并发的特性，特别是所有的执行流都在一个进程内部，可以共享该进程所拥有的资源(代码、数据及堆栈)。这种比进程粒度还小的执行单位就是线程。如果图 3-1 中的 Web 服务器进程是多线程的，那么这种服务器可以创建一个单独的线程，以便监听客户请求。当有请求时，Web 服务器不是创建进程，而是创建线程以处理请求，并恢复监听其他请求，如图 3-4 所示。

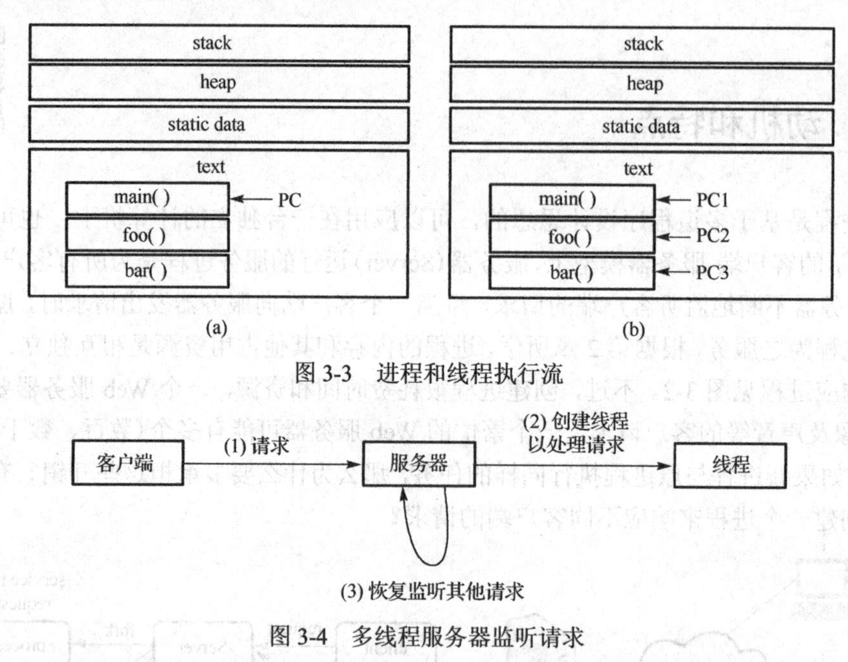

图 3-3　进程和线程执行流

图 3-4　多线程服务器监听请求

3.2　线程定义

线程是操作系统能够进行计算调度的最小单位。大部分情况下，它包含在进程之中，是进程中的实际执行单位。一个线程是指进程中一个单一顺序的控制流，一个进程中可以并发多个线程，每个线程并行执行不同任务。在 UNIX System V 及 Sun OS 中，线程也称为轻量进程(Lightweight Processes)，但轻量进程更多指内核线程(Kernel Thread)，而用户线程(User Thread)仍称为线程。

如图 3-5 所示，同一进程中的多个线程将共享该进程中的全部系统资源，如虚拟地址空间、文件描述符等。但同一进程中的多个线程有各自的调用栈，自己的寄存器环境。如果进程要完成的任务很多，就需要很多线程，也要调用很多核心，在多核、多 CPU 或支持 Hyper-threading 的 CPU 上使用多线程程序的好处是显而易见的，即提高了程序运行吞吐率。以人工作的样子想象，核心相当于人，人越多则能同时处理的事情越多，而线程相当于手，手越多则工作效率越高。在单 CPU 单核的计算机上，使用多线程技术，也可以把进程中负责 I/O 处理、人机交互而常被阻塞的部分与密集计算的部分分开运行，编写专门的 workhorse 线程执行密集计算。虽然多任务比不上多核，但因为具备多线程的能力，从而提高了程序的运行效率。

采用多线程的优点如下。

(1) 响应性。

(2) 资源共享。

(3) 经济。

(4) 可伸缩性。

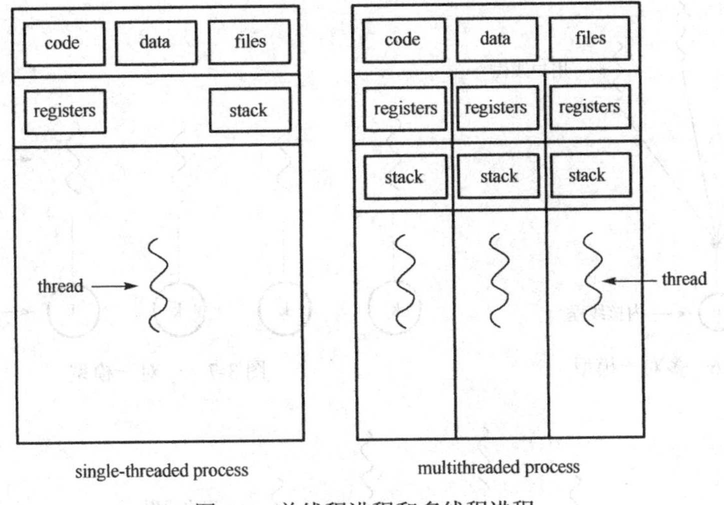

图 3-5　单线程进程和多线程进程

3.3 线程模型

有两种方法能提供线程支持：用户层的用户线程(User Thread)及内核层的内核线程(Kernel Thread)。用户线程位于内核之上，对它的管理无须内核支持；而内核线程由操作系统直接支持与管理。几乎所有的现代操作系统，包括 Windows、Linux、macOS 及 Solaris，都支持内核线程。

用户线程和内核线程之间必然存在某种关系。下面研究三种常用的建立这种关系的方法：多对一模型、一对一模型和多对多模型。

1. 多对一模型

多对一模型(如图 3-6 所示)映射多个用户线程到一个内核线程。线程管理是由用户空间的线程库来完成的，因此效率比较高。不过，用户线程的调度是在用户模式下完成的，系统内核调度的单位是内核线程，因此当一个用户线程阻塞时，整个内核线程也会阻塞，直接的后果就是所有用户线程全部阻塞。此外，因为任意时间只有一个线程可以访问内核，所以多个线程不能并发运行在多处理器系统上，因此现在几乎没有操作系统继续使用这个模型。

2. 一对一模型

一对一模型(如图 3-7 所示)映射一个用户线程到一个内核线程。该模型在一个线程阻塞时，能够允许另一个线程继续运行，所以它提供了比多对一模型更好的并发功能；同时也允许多个线程并发运行在多处理器系统上。这种模型的缺点是，创建一个用户线程就要创建一个相应的内核线程。由于创建内核线程的开销会影响应用程序的性能，所以这种模型的实现限制了系统支持的线程数量。Linux 和 Windows 系统都实现了一对一模型。

3. 多对多模型

多对多模型(如图 3-8 所示)多路复用多个用户线程到同样数量或更少的内核线程中。内核线程的数量可能与特定应用程序或特定机器有关(应用程序在多处理器上比在单处理器上可能被分配到更多的线程)。

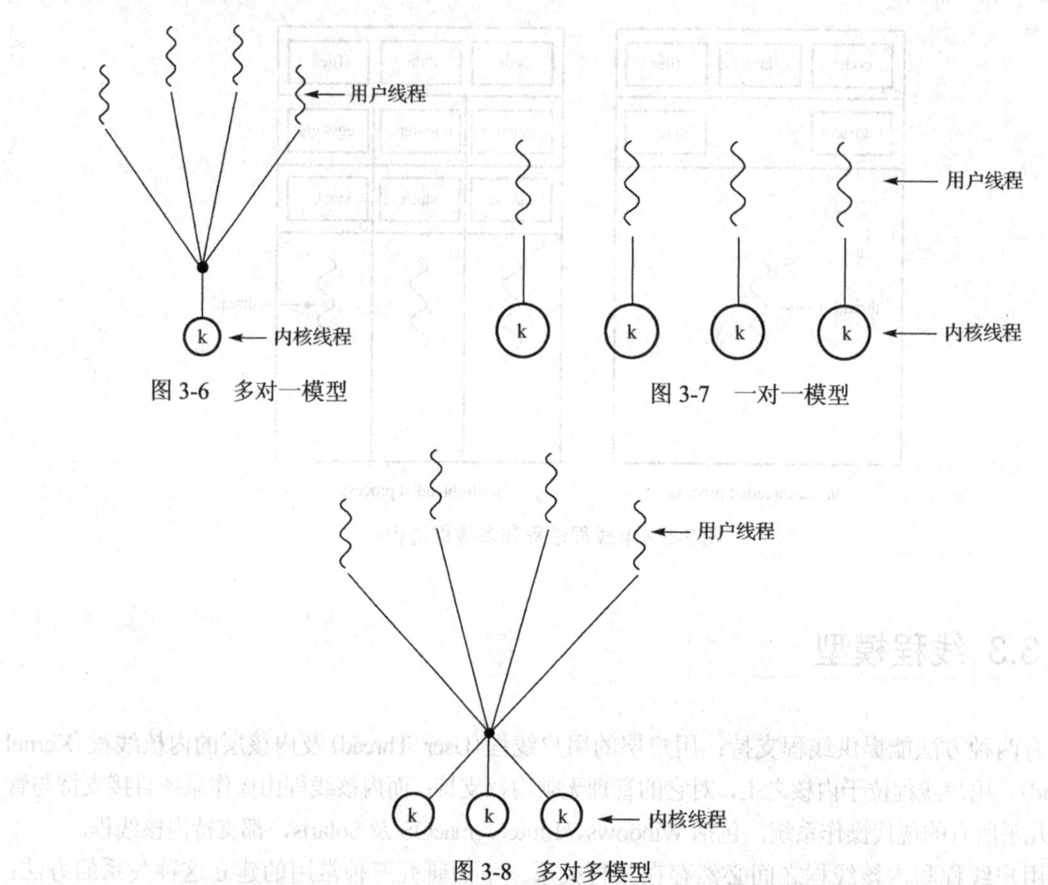

图 3-6　多对一模型

图 3-7　一对一模型

图 3-8　多对多模型

下面分析这些设计对并发性的影响。虽然多对一模型允许开发人员创建任意多的用户线程，但是由于内核只能一次调度一个线程，所以并未增加并发性。虽然一对一模型提供了更好的并发性，但是开发人员应注意不要在应用程序内创建太多线程(有时系统可能会限制创建线程的数量)。多对多模型没有以上两个缺点，开发人员可以创建任意多的用户线程，并且相应的内核线程能在多处理器系统上并发运行。此外，当一个线程阻塞时，内核还可以调度另一个线程来运行。

3.4 线程库

线程库(Thread Library)为开发人员提供创建及管理线程的API。实现线程库的主要方法有两种：第一种是在用户空间中提供一个没有内核支持的库，这种库的所有代码和数据结构都位于用户空间中，这意味着调用库内的一个函数只导致用户空间内一个本地函数的调用，而不会导致系统调用；第二种是实现由操作系统直接支持的内核级的一个库。对于这种情况，库内的代码和数据结构位于内核空间，调用库中的一个API函数通常会导致对内核的系统调用。

目前常用的三种线程库是 POSIX Thread(简称 Pthread)、Windows Thread 及 Java

Thread。Pthread 是 POSIX 标准定义的线程创建与同步 API，是线程行为的规范(Specification)，而不是实现(Implementation)。操作系统开发人员可以根据意愿采取任何形式来实现。许多操作系统都实现了这个线程规范，大多数为UNIX类型的系统，如Linux、macOS 及 Solaris。虽然 Windows 本身并不支持 Pthread，但是有些第三方为 Windows 提供了 Pthread 的实现。

Windows Thread 是用于 Windows 系统的内核级线程库。Java 线程 API 允许在 Java 程序中直接创建和管理线程。然而，由于大多数 JVM 实例运行在宿主操作系统上，Java 线程 API 通常采用宿主系统的线程库来实现。这意味着 Java 线程在 Windows 系统上通常采用 Windows API 来实现，而在 UNIX 及 Linux 系统中采用 Pthread 来实现。

本章 3.6 节的实验 3 会指导读者使用 Pthread 线程库创建线程，只有在实践过程中才能深入体会到线程和进程之间的差异。因为线程在操作上更为灵活，学习线程更加便于对后续章节知识的理解，从本章实验开始，后续的所有实验部分均以线程为调度单位，在本书中，除特别说明，“并发进程”和“并发线程”，“进程调度”和“线程调度”指代同一个意思。

3.5 Reading Materials

3.5.1 Overview

A thread is a flow of control within a process. A multithreaded process contains several different flows of control within the same address space. The benefits of multithreading include increased responsiveness to the user, resource sharing within the process, economy, and scalability factors, such as more efficient use of multiple processing cores.

User-level threads are threads that are visible to the programmer and are unknown to the kernel. The operating-system kernel supports and manages kernel-level threads. In general, user-level threads are faster to create and manage than are kernel threads, because no intervention from the kernel is required.

Three different types of models relate user and kernel threads. The many-to-one model maps many user threads to a single kernel thread. The one-to-one model maps each user thread to a corresponding kernel thread. The many-to-many model multiplexes many user threads to a smaller or equal number of kernel threads.

Most modern operating systems provide kernel support for threads. These include Windows, Mac OS, Linux, and Solaris.

Thread libraries provide the application programmer with an API for creating and managing threads. Three primary thread libraries are in common use: Pthreads, Windows threads, and Java threads.

In addition to explicitly creating threads using the API provided by a library, we can use implicit threading, in which the creation and management of threading is transferred to compilers and run-time libraries. Strategies for implicit threading include thread pools, OpenMP, and Grand Central Dispatch.

Multithreaded programs introduce many challenges for programmers, including the semantics of the fork() and exec() system calls. Other issues include signal handling, thread cancellation, thread-local storage, and scheduler activations.

3.5.2 POSIX Thread（Pthread）Libraries

The POSIX thread libraries are a standards based thread API for C/C++. It allows one to spawn a new concurrent process flow. It is most effective on multi-processor or multi-core systems where the process flow can be scheduled to run on another processor thus gaining speed through parallel or distributed processing. Threads require less overhead than "forking" or spawning a new process because the system does not initialize a new system virtual memory space and environment for the process. While most effective on a multiprocessor system, gains are also found on uniprocessor systems which exploit latency in I/O and other system functions which may halt process execution. (One thread may execute while another is waiting for I/O or some other system latency.) Parallel programming technologies such as MPI and PVM are used in a distributed computing environment while threads are limited to a single computer system. All threads within a process share the same address space. A thread is spawned by defining a function and it's arguments which will be processed in the thread. The purpose of using the POSIX thread library in your software is to execute software faster.

Thread Basics:

- Thread operations include thread creation, termination, synchronization (joins,blocking), scheduling, data management and process interaction.
- A thread does not maintain a list of created threads, nor does it know the thread that created it.
- All threads within a process share the same address space.
- Threads in the same process share:
  - Process instructions
  - Most data
  - open files (descriptors)
  - signals and signal handlers
  - current working directory
  - User and group id
- Each thread has a unique:
  - Thread ID
  - set of registers, stack pointer
  - stack for local variables, return addresses
  - signal mask
  - priority
  - Return value: errno
- pthread functions return "0" if OK.

3.6 实验3 Pthread 多线程

获取视频

1. 实验目的

通过编码熟悉 Pthread 线程库的基本使用方法，观察多线程的运行特点，对比第2章并发进程实验，发现并发进程和并发线程之间的差异，最后使用多线程完成一个订票系统。

2. 实验方法

自行学习、检索 Pthread 库的函数说明，按照实验内容完成实验，有困难的读者可以在视频指导下完成。

3. 实验内容1

采用多线程机制，使用 Monte Carlo 技术圆周率的值。该实验内容是让读者了解在不同数量的 CPU 核心条件下，多线程的工作性能有什么变化。请读者自行阅读相关资料，在视频的指导下完成。

4. 实验内容2

(1) 创建线程。创建一个新的源文件 gedit helloThread.c，并输入下面的代码。

```
 1 #include <sys/types.h>
 2 #include <unistd.h>
 3 #include <stdio.h>
 4 #include <pthread.h>
 5 void* threadFunc(void* arg){ //线程函数
 6     printf("In NEW thread\n");
 7 }
 8 int main()
 9 {
10     pthread_t tid;
11     pthread_create(&tid, NULL, threadFunc, NULL);
12     pthread_join(tid, NULL);
13     printf("In main thread\n");
14     return 0;
15 }
```

编译这段程序时，注意 gcc 指令中要加入新的参数，指令如下：

```
gcc helloThread.c -o helloThread -l pthread
```

运行程序，观察到了什么现象？为什么？

(2) 试着在主线程和新线程中加入循环输出，观察输出的效果和并发父子进程的运行效果是否相似。

(3) 多线程订票系统示例代码如下。

```
1 #include <stdio.h>
2 #include <pthread.h>
3 #include <unistd.h>
4 int ticketAmount = 2; //Global Variable
```

```
 5 void* ticketAgent(void* arg){
 6   int t = ticketAmount;
 7   sleep(1);
 8   if (t > 0)
 9    {
10      printf("One ticket sold!\n");
11      t--;
12   }else{
13      printf("Ticket sold out!!\n");
14    }
15   ticketAmount = t;
16   pthread_exit(0);
17 }
18
19 int main(int argc, char const *argv[])
20 {
21   pthread_t ticketAgent_tid[2];
22   for (int i = 0; i < 2; ++i)
23    {
24      pthread_create(ticketAgent_tid+i, NULL, ticketAgent,NULL);
25    }
26   for (int i = 0; i < 2; ++i)
27    {
28      pthread_join(ticketAgent_tid[i],NULL);
29    }
30   printf("The left ticket is %d\n", ticketAmount);
31   return 0;
32 }
```


以上代码第 22～25 行用一个线程函数创建了两个线程，这种创建方法使用循环来减少代码的书写量，第 26～29 行给出了如何用循环来等待所有线程结束。两个线程作为订票系统，共享一个全局变量 ticketAmount，两个线程运行结束后，在主线程中打印余票量，编译及运行这个程序，观察运行结果。

第4章　进程调度

获取视频

4.1　概述

当计算机系统是多道批处理系统时，通常会有多个进程或线程同时竞争 CPU。只要有两个或更多的进程就绪，那这种情形就会发生。如果只有一个 CPU 可用，那么就必须选择下一个要运行的进程。在操作系统中，完成选择工作的这部分程序称为调度器(Scheduler)，该程序使用的算法称为调度算法(Scheduling Algorithm)。调度器可能会针对不同的目标设计，如吞吐率最大化、响应时间最小化及最低延迟或最大化公平。但在实践中，这些目标通常是互相冲突的，因此，调度器会实现一个权衡利弊的折中方案。

操作系统一般提供三个级别的进程调度：长程调度(Long-Term Scheduling，LTS)、中程调度(Medium-Term Scheduling，MTS)及短程调度(Short-Term Scheduling，STS)。

1. 长程调度

该类调度决定了任务或进程是否能插入就绪队列(内存中)，当系统尝试运行一个程序时，调度器会根据系统资源情况决定加载其进内存或是延迟运行。调度器掌控着谁能在系统上运行，同时还决定了并发的程度，即同时运行程序的数量。此外，调度器会对 I/O 密集型(I/O-bound)和计算密集型(CPU-bound)进程做出划分。

在图 4-1(a)中，进程花费了绝大多数时间在计算上，而图 4-1(b)所示的进程则在等待 I/O 上花费了绝大多数时间。前者称为计算密集型进程或 CPU 密集型进程，后者称为 I/O 密集型进程。典型的计算密集型进程具有较长时间的 CPU 集中使用和较小频度的 I/O 等待；典型的 I/O 密集型进程具有较短时间的 CPU 集中使用和频繁的 I/O 等待。

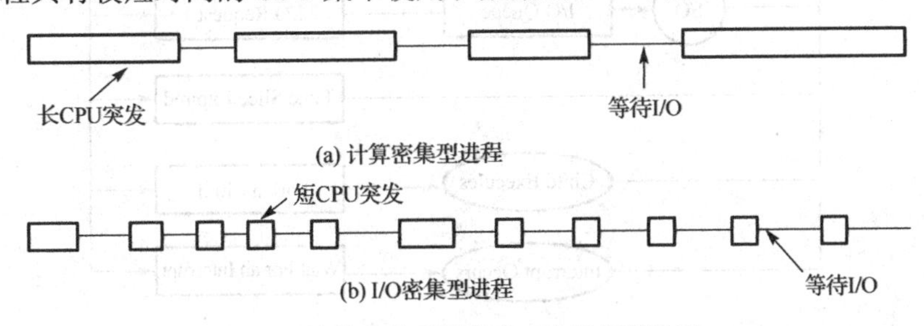

图 4-1　计算密集型进程和 I/O 密集型进程

随着 CPU 的计算速度越来越快，更多的进程倾向于 I/O 密集型进程，因为 CPU 计算速度的改进比磁盘快得多，对 I/O 密集型进程的调度处理更为重要。如果需要运行 I/O 密集型进程，那么就应该让它尽快得到机会，以便发出 I/O 请求并保持磁盘始终忙碌。从图 4-1 中可以看到，如果进程是 I/O 密集型的，将需要多运行一些这类进程以保持对 CPU 的充分利用。

因此，选出一个 I/O 密集型进程和计算密集型进程的良好组合，对于长程调度器是非常重要的。否则，假如所有的进程都是计算密集型的，那么 I/O 队列将几乎永远是空的，这样就会导致一些设备从未被使用过，从而导致系统资源分配不均衡。显然，性能极佳的系统必然是计算密集型进程和 I/O 密集型进程的组合。在现代操作系统中，这被用来保证实时进程能获得足够的 CPU 时间来完成任务。

2. 中程调度

中程调度会临时将那些一直不活跃的进程、优先级低的进程、频繁产生错误的进程或者占用大量内存的进程从内存中移出，放入第二级存储设备(Secondary Storage)的交换区中，以腾出更多的内存资源。当系统内存充足时，或者程序不再处于等待态时，调度器又会将刚刚被放入交换区中的进程重新放入内存中。这两个动作通常称为“换出”(Swap Out)和“换入”(Swap In)。

3. 短程调度

短程调度(CPU 调度)决定了在一个时钟中断、I/O 中断及系统调用其他种类的信号之后，应该运行内存中的哪个进程，将 CPU 分配给谁。这种调度可以是抢占式的(Preemptive Scheduling)，能够强行把一个在 CPU 中运行的进程中断，然后分配给其他进程；也可以是非抢占式的(Non Preemptive Scheduling)，无法强行把进程从 CPU 上中断。

第 2 章介绍过操作系统是通过 PCB 来管理进程实体的，进程的各种状态也记录在 PCB 中，操作系统将相同状态的进程 PCB 以队列(先进先出)的形式管理起来，因为处于等待态的进程会比较多，所以按照等待事件的不同形成了若干等待队列。图 4-2 中的矩形框表示 PCB 队列，就绪队列只有一个，而 I/O 请求、时间片过期、子进程创建及等待中断是四个等待队列，进程的状态迁移从另一个角度来看也是其 PCB 在不同队列中移出和插入的过程。

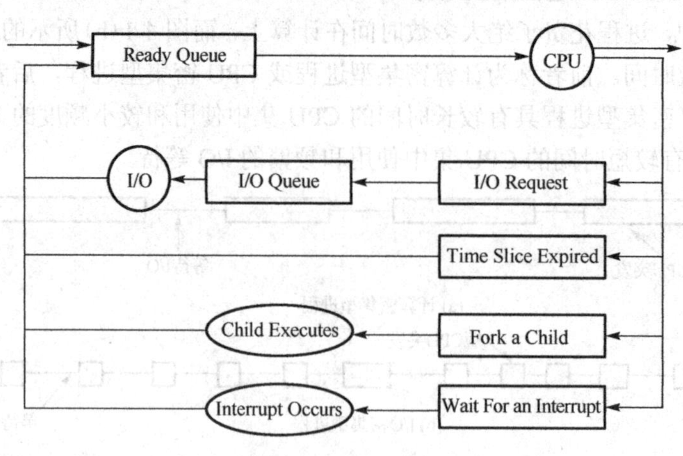

图 4-2 进程队列

图 4-3 完整地展示了三个级别的进程调度，长程调度决定哪个程序可以加载进内存，插入就绪队列，中程调度将进程换出至第二级存储设备并视机换入就绪队列，短程调度从就绪队列中挑选一个进程分配给其 CPU，在这之前按照抢占或非抢占的行为决定是否将正在 CPU 上运行的进程踢出去(中断)。

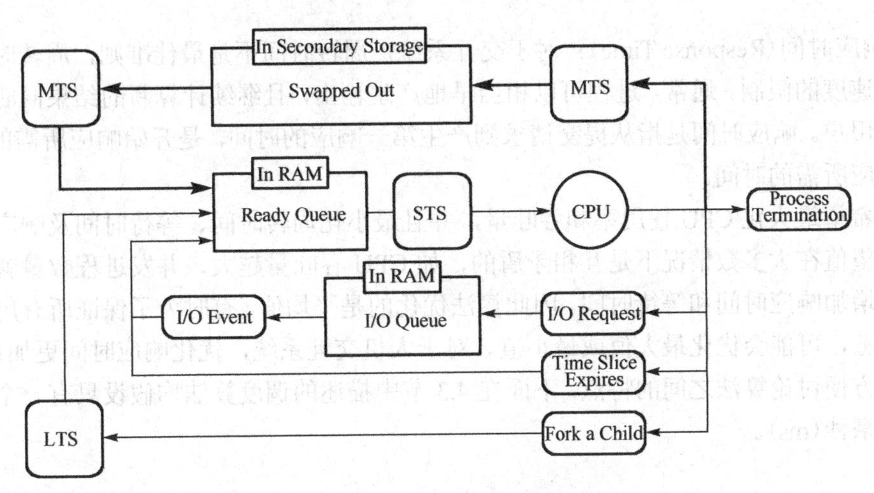

图 4-3　进程调度级别

在三个级别的调度中，短程调度的频率是最高的。本章的重点就是短程调度。虽然调度算法大多是通用的，但短程调度算法因其工作特点也有其特殊的地方。同时，适用于进程调度的处理方法也同样适用于线程调度。当内核管理线程的时候，调度经常是按线程级别的，与线程所属的进程基本没有关联。

调度器(Dispatcher)是操作系统的一个模块，其功能如下。

(1)切换进程上下文。

(2)切换到用户模式。

(3)跳转到用户程序的合适位置，以便重新启动程序。

调度程序的运行速度应尽可能地快，因为在每次进程切换时都要运行调度程序。调度程序中断一个进程而运行另一个进程所需的时间称为调度延迟(Dispatch Latency)。

4.2　调度标准

不同的调度算法具有不同的属性，选择一个特定的算法会对某些进程更有利。为了比较调度算法，可以采用多种比较准则。选择哪些准则来做比较，对于确定哪种算法是最好的至关重要。这些比较准则如下。

(1)CPU 使用率(CPU Utilization)：应使 CPU 尽可能地忙碌。从概念上来说，CPU 使用率的范围是从 0%到 100%。但对于一个实际的系统，它的范围是从 40%(轻负荷系统)到 90%(重负荷系统)。

(2)吞吐量(Throughput)：指在一个时间单元内进程完成的数量。对于长进程，吞吐量可能为每小时一个进程；对于短进程，吞吐量可能为每秒十个进程。

(3)周转时间(Turnaround Time)：从一个特定进程的角度来看，一个重要的准则是运行这个进程需要多长时间。从进程提交请求到进程完成的时间段称为周转时间。周转时间包括等待进入内存、在就绪队列中等待、在 CPU 上运行及 I/O 操作的时间之和。

(4)等待时间(Waiting Time)：CPU 调度算法并不影响进程运行和 I/O 操作的时间，只影响进程在就绪队列中等待所需的时间。等待时间是指进程在就绪队列中等待所花费的时间。

(5) 响应时间(Response Time)：对于交互系统，周转时间不是最佳准则，周转时间通常受输出设备速度的限制。通常，进程可以相当早地产生输出，且继续计算新的结果同时输出以前的结果给用户。响应时间是指从提交请求到产生第一响应的时间，是开始响应所需的时间，而非输出响应所需的时间。

我们希望最大化CPU使用率和吞吐量，并且最小化周转时间、等待时间及响应时间。但是这些期望值在大多数情况下是互相矛盾的，如CPU吞吐量越大，并发进程数量就会越多，这势必会增加响应时间和等待时间。因此算法优化的是平均值，有时为了保证所有用户都能得到好的服务，可能会优化最大值或最小值。对于人机交互系统，优化响应时间更加重要。

为了方便讨论算法之间的特点，下面在4.3节中描述的调度算法均假设只有一个CPU，时间单位为毫秒(ms)。

4.3 调度算法

4.3.1 先来先服务调度

最简单的调度算法是先来先服务(First-Come First-Served，FCFS)调度算法。采用这种算法，先请求CPU的进程首先被分配到CPU。FCFS算法可以通过FIFO队列容易地实现。当一个进程进入就绪队列时，它的PCB会被链接到队列尾部。当CPU空闲时，它会被分配给位于队列头部的进程，并且把这个进程从队列中移去。FCFS算法是非抢占式的。一旦CPU被分配给了一个进程，该进程就会一直使用CPU，直到释放为止(程序终止或请求I/O)。

FCFS算法的缺点是平均等待时间很长。假设有如下一组进程，它们在0时刻到达，CPU执行时间按ms计：

进程	CPU执行时间
P1	28
P2	9
P3	3

如果进程按P1、P2及P3的顺序到达，并且按FCFS算法处理，那么可以得到如图4-4所示的调度图，图中的数字表示每个进程的开始与结束时间。

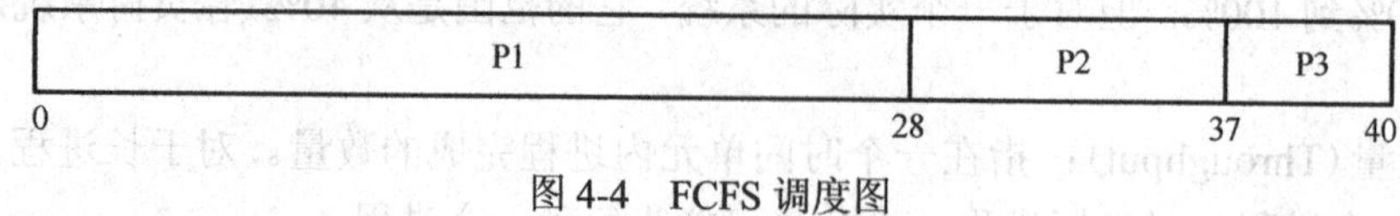

图4-4 FCFS调度图

进程P1的等待时间为0 ms，进程P2的等待时间为28ms，而进程P3的等待时间为37ms。因此，平均等待时间为(0+28+37)/3=22ms。进程P1的周转时间是28ms，进程P2的周转时间是37ms，进程P3的周转时间是40ms，三个进程的平均周转时间为(28+37+40)/3=35ms。

不过，如果进程按P3、P2、P1的顺序到达，那么结果如图4-5所示。

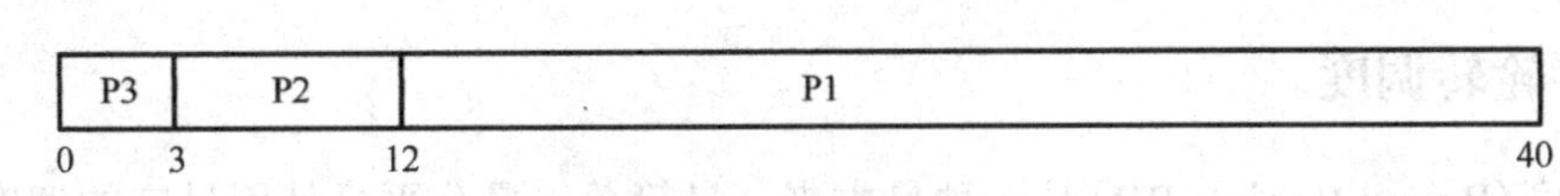

图 4-5　FCFS 调度图

此时平均等待时间为(12+3+0)/3=5ms，平均周转时间为(3+12+40)/3=18ms，可见等待时间和周转时间大大减少。因此，FCFS 算法的平均等待时间通常不是最小的，而且如果进程的 CPU 执行时间变化很大，那么平均等待时间的变化也会很大。

从这个现象我们发现如果调度短进程优先运行，会极大地降低平均等待时间和平均周转时间，依据这个现象就有了最短作业优先调度算法。

4.3.2　最短作业优先调度

最短作业优先(Shortest Job First，SJF)调度算法将每个进程与其下一次 CPU 执行时间关联起来。当 CPU 空闲时，它会被赋给具有最短 CPU 执行时间的进程。如果两个进程具有同样的 CPU 执行时间，那么再按 FCFS 算法处理。所以，对 SJF 算法更恰当的定义是“最短下次 CPU 执行”(Shortest Next CPU Burst)算法，这是因为调度取决于进程的下一次 CPU 执行时间，而不是其总的执行时间。作为 SJF 调度算法的例子，假设有如下一组进程，CPU 执行时间以 ms 计：

进程	CPU 执行时间
P1	6
P2	8
P3	7
P4	3

采用 SJF 调度算法，根据图 4-6 所示的调度图来调度这些进程。

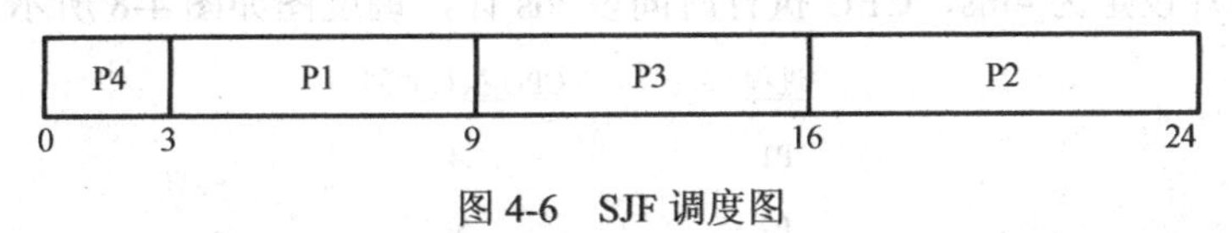

图 4-6　SJF 调度图

进程 P1 的等待时间为 3ms，进程 P2 的等待时间为 16ms，进程 P3 的等待时间为 9ms，进程 P4 的等待时间为 0ms。因此，平均等待时间为(3+16+9+0)/4=7ms，平均周转时间为 13ms。如果采用 FCFS 算法，那么平均等待时间为 10.25ms，平均周转时间为 16.25ms。

SJF 算法被证明是最优的算法，这是因为对于给定的一组进程，SJF 算法的平均等待时间最少。通过将短进程移到长进程之前，短进程等待时间的减少大于长进程等待时间的增加。因此，平均等待时间减少。

但遗憾的是，SJF 算法是很难实现的，因为我们没有办法知道进程下一次所需要的 CPU 执行时间。虽然有很多解决方案，比如让用户提交作业时就估计所需的时间，但用户可能会为了获得优先权故意低估；再如系统对进程所需要的下次 CPU 执行时间进行算法预测来近似实现 SJF 算法。

4.3.3 轮转调度

轮转调度(Round Robin，RR)是一种最古老、最简单、最公平且使用最广的调度算法。每个进程被分配一个时间段，称为时间片(Quantum)，即允许该进程在该时间片中运行。轮转调度算法是抢占式的，若在时间片结束时该进程还在运行，则将剥夺其CPU使用权并将CPU分配给其他进程。若该进程在时间片结束前阻塞或结束，则立即进行CPU切换。轮转调度很容易实现，调度程序所要做的就是维护一个可运行进程列表[如图 4-7(a)所示]，当一个进程用完它的时间片后，就被移到队列的末尾，如图 4-7(b)所示。

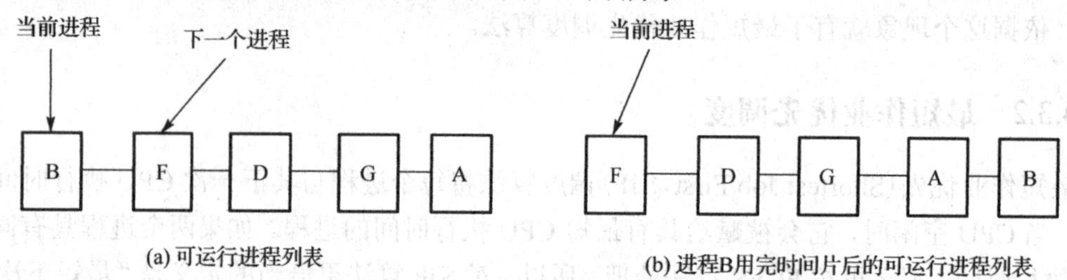

图 4-7 轮转调度

轮转调度中时间片长度的选择非常重要，过长或过短都会影响系统资源利用率。每次时间片用尽都会发生进程切换，假设一次进程切换大约消耗1ms的CPU执行时间，那么100ms时间片将花费1%的时间管理切换开销，50ms时间片的切换开销为2%，4ms时间片的切换开销为 20%。由此可见，时间片越小，切换开销比例会越大；但时间片增大，会增加进程的平均响应时间。考虑一种极端情况，即时间片无穷大，那么轮转调度将退化成FCFS调度。大多数现代操作系统的时间片为10～100 ms，上下文切换的时间一般小于10 ms。

轮转调度的平均等待时间通常较长，因为长进程可能需要多个时间片才能完成计算，每当它用完一个时间片都不得不等待下一次时间片的分配。例如，有如下一组进程，它们都在0时刻到达，时间片设定为4ms，CPU执行时间以ms计，调度图如图 4-8 所示。

进程	CPU 执行时间
P1	24
P2	3
P3	3

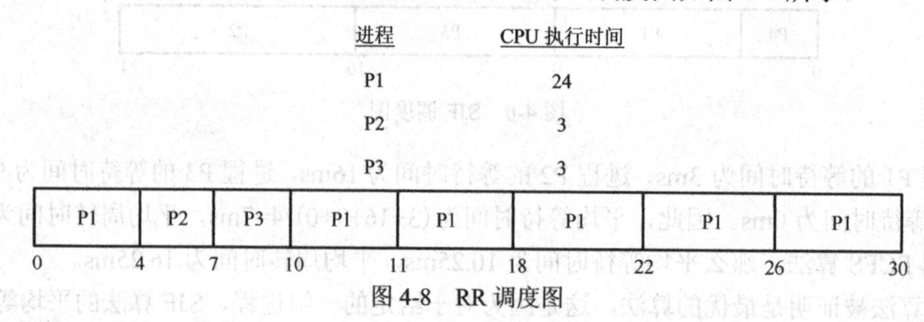

图 4-8 RR 调度图

如果使用4 ms的时间片，那么P1会执行最初的4 ms。由于它还需要20 ms，所以在第一个时间片之后它会被抢占，而CPU则会被分配给队列中的下一个进程。由于P2的执行不需要4 ms，所以在其时间片用完之前就会退出。该例中轮转调度的平均等待时间为(6+4+7)/3=5.66 ms。

最后，为了更直观地比较FCFS、SJF和RR三种调度算法的特点，我们对相同的一组数据进行模拟调度，从调度图中体会一下各自的优缺点。该例中有5个进程，都在0时刻到达，时间片为1000ms，每个进程的CPU执行时间如下(单位 ms)，FCFS、SJF和RR调度图分别如图 4-9、图 4-10 及图 4-11 所示。

进程	CPU 执行时间
P1	1000
P2	2000
P3	3000
P4	4000
P5	5000

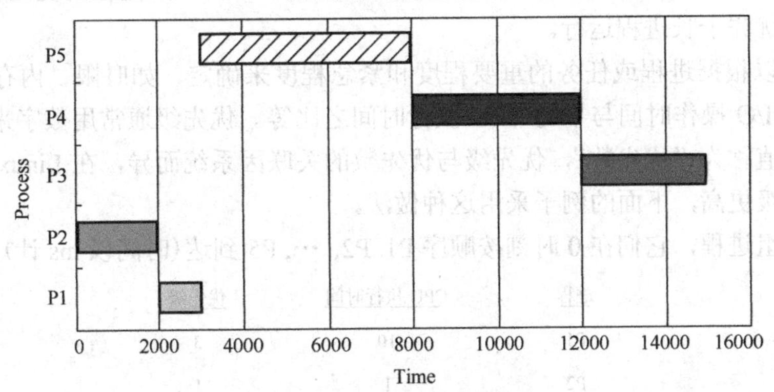

图 4-9　FCFS 调度图

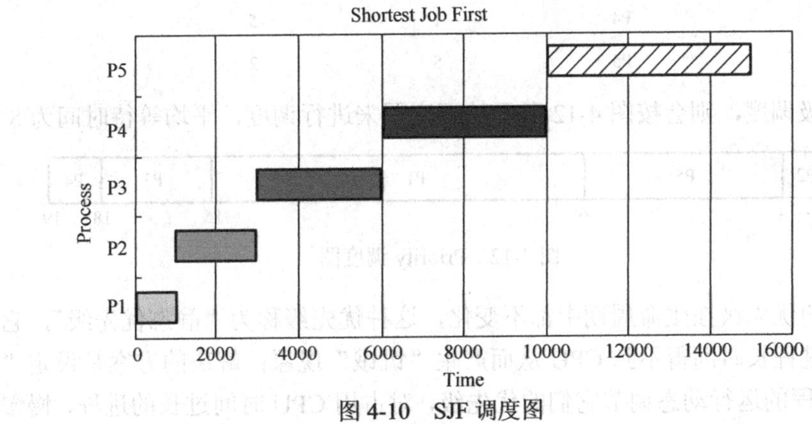

图 4-10　SJF 调度图

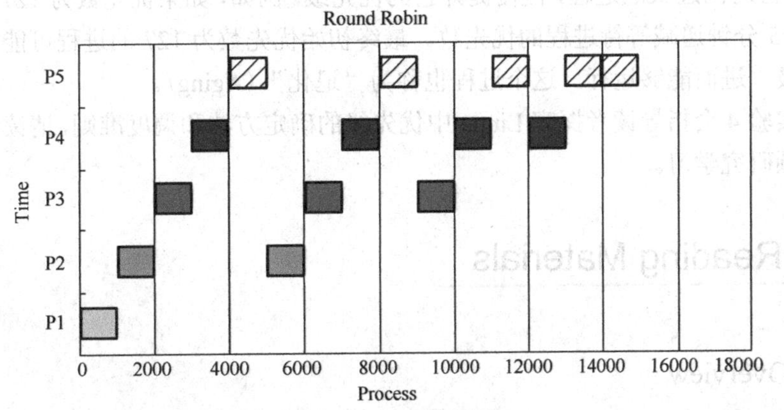

图 4-11　RR 调度图

4.3.4 优先级调度

优先级调度(Priority Scheduling)算法根据确定的优先级选取下一个运行的进程，总选择就绪队列中优先级最高的进程投入运行。优先级调度可以是抢占式的或非抢占式的。当一个进程到达就绪队列时，比较它的优先级与当前运行进程的优先级，如果新进程的优先级高于当前运行进程的优先级，那么抢占优先级调度算法就会抢占 CPU，非抢占优先级调度算法只将新的进程加到就绪队列的头部。SJF 是优先级调度算法的一个特例，短进程的优先级比长进程高，所以短进程会优先于长进程运行。

优先级可以根据进程或任务的重要程度和紧急程度来确定，如时限、内存要求、打开文件数量及平均 I/O 操作时间与平均 CPU 执行时间之比等。优先级通常用数字来表示，用于表示优先级的数值称为“优先数”，优先级与优先数的关联因系统而异，在 Linux 系统中，优先数小代表优先级更高，下面的例子采用这种做法。

有如下一组进程，它们在 0 时刻按顺序 P1, P2, …, P5 到达(时间以 ms 计)：

进程	CPU 执行时间	优先级
P1	10	3
P2	1	1
P3	2	4
P4	1	5
P5	5	2

采用优先级调度，则会按图 4-12 所示的调度图来进行调度，平均等待时间为 8.2ms。

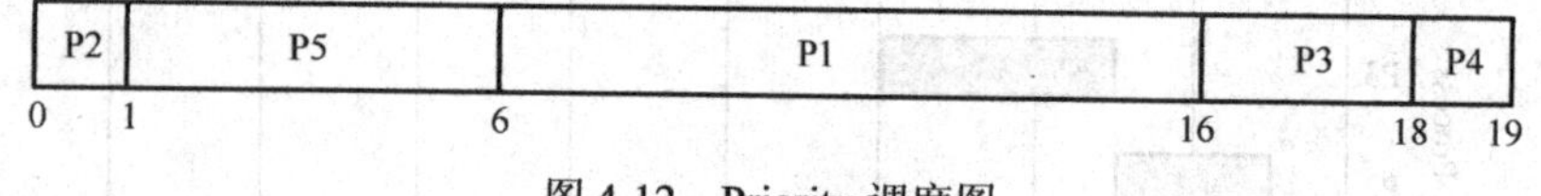

图 4-12　Priority 调度图

如果进程的优先级在生命周期中都不变化，这种优先级称为“静态优先级”，它可能会导致低优先级的进程长时间得不到 CPU 从而产生“饥饿”现象；解决的方案是设定“动态优先级”，即随着进程的运行动态调节它们的优先级，对占用 CPU 时间过长的进程，慢慢降低它的优先级，对等待时间过长的进程，慢慢提升它的优先级。例如，如果优先数为 127(低)～0(高)，那么可以每 15 分钟递减等待进程的优先数。最终初始优先数为 127 的进程可能会拥有系统内最高的优先级，进而能够运行，这个过程也称为“退化”(Aging)。

本章的实验 4 会指导读者探究 Linux 中优先级的确定方法和调度准则，请读者参照实验指导、观看视频研究学习。

4.4 Reading Materials

4.4.1 Overview

CPU scheduling is the task of selecting a waiting process from the ready queue and allocating the CPU to it. The CPU is allocated to the selected process by the dispatcher.

First-Come, First-served (FCFS) scheduling is the simplest scheduling algorithm, but it can cause short processes to wait for very long processes. Shortest Job First (SJF) scheduling is provably optimal, providing the shortest average waiting time. Implementing SJF scheduling is difficult, however, because predicting the length of the next CPU burst is difficult. The SJF algorithm is a special case of the general priority scheduling algorithm, which simply allocates the CPU to the highest-priority process. Both priority and SJF scheduling may suffer from starvation. Aging is a technique to prevent starvation.

Round Robin (RR) scheduling is more appropriate for a time-shared (interactive) system. RR scheduling allocates the CPU to the first process in the ready queue for q time units, where q is the time quantum. After q time units, if the process has not relinquished the CPU, it is preempted, and the process is put at the tail of the ready queue. The major problem is the selection of the time quantum. If the quantum is too large, RR scheduling degenerates to FCFS scheduling. If the quantum is too small, scheduling overhead in the form of context-switch time becomes excessive.

The FCFS algorithm is nonpreemptive; the RR algorithm is preemptive; The SJF and priority algorithms may be either preemptive or nonpreemptive.

Multilevel queue algorithms allow different algorithms to be used for different classes of processes. The most common model includes a foreground interactive queue that uses RR scheduling and a background batch queue that uses FCFS scheduling. Multilevel feedback queues allow processes to move from one queue to another.

Many contemporary computer systems support multiple processors and allow each processor to schedule itself independently. Typically, each processor maintains its own private queue of processes (or threads), all of which are available to run. Additional issues related to multiprocessor scheduling include processor affinity, load balancing, and multicore processing.

A real-time computer system requires that results arrive within a deadline period; results arriving after the deadline has passed are useless. Hard real-time systems must guarantee that real-time tasks are serviced within their deadline periods. Soft real-time systems are less restrictive, assigning real-time tasks higher scheduling priority than other tasks.

Real-time scheduling algorithms include rate-monotonic and earliest deadline-first scheduling. Rate-monotonic scheduling assigns tasks that require the CPU more often a higher priority than tasks that require the CPU less often. Earliest-deadline-first scheduling assigns priority according to upcoming deadlines—the earlier the deadline, the higher the priority. Proportional share scheduling divides up processor time into shares and assigning each process a number of shares, thus guaranteeing each process a proportional share of CPU time. The POSIX Pthread API provides various features for scheduling real-time threads as well.

Operating systems supporting threads at the kernel level must schedule threads—not processes—for execution. This is the case with Solaris and Windows. Both of these systems schedule threads using preemptive, priority based scheduling algorithms, including support for real-time threads. The Linux process scheduler uses a priority-based algorithm with real-time support as well. The scheduling algorithms for these three operating systems typically favor interactive over CPU-bound processes.

The wide variety of scheduling algorithms demands that we have methods to select among algorithms. Analytic methods use mathematical analysis to determine the performance of an algorithm. Simulation methods determine performance by imitating the scheduling algorithm on a "representative" sample of processes and computing the resulting performance. However, simulation can at best provide an approximation of actual system performance. The only reliable technique for evaluating a scheduling algorithm is to implement the algorithm on an actual system and monitor its performance in a "real-world" environment.

4.4.2 CFS: Completely Fair Process Scheduling in Linux

Linux takes a modular approach to processor scheduling in that different algorithms can be used to schedule different process types. A scheduling class specifies which scheduling policy applies to which type of process. Completely fair scheduling (CFS), which became part of the Linux 2.6.23 kernel in 2007, is the scheduling class for normal (as opposed to real-time) processes and therefore is named SCHED_NORMAL.

CFS is geared for the interactive applications typical in a desktop environment, but it can be configured as SCHED_BATCH to favor the batch workloads common, for example, on a high-volume web server. In any case, CFS breaks dramatically with what might be called "classic preemptive scheduling." Also, the "completely fair" claim has to be seen with a technical eye; otherwise, the claim might seem like an empty boast.

Let's dig into the details of what sets CFS apart from—indeed, above—other process schedulers. Let's start with a quick review of some core technical terms.

Linux inherits the Unix view of a process as a program in execution. As such, a process must contend with other processes for shared system resources: memory to hold instructions and data, at least one processor to execute instructions, and I/O devices to interact with the external world. Process scheduling is how the operating system (OS) assigns tasks (e.g., crunching some numbers, copying a file) to processors—a running process then performs the task. A process has one or more threads of execution, which are sequences of machine-level instructions. To schedule a process is to schedule one of its threads on a processor.

In a simplifying move, Linux turns process scheduling into thread scheduling by treating a scheduled process as if it were single-threaded. If a process is multi-threaded with N threads, then N scheduling actions would be required to cover the threads. Threads within a multi-threaded process remain related in that they share resources such as memory address space. Linux threads are sometimes described as lightweight processes, with the lightweight underscoring the sharing of resources among the threads within a process.

Although a process can be in various states, two are of particular interest in scheduling. A blocked process is awaiting the completion of some event such as an I/O event. The process can resume execution only after the event completes. A runnable process is one that is not currently blocked.

A process is processor-bound (aka compute-bound) if it consumes mostly processor as opposed to I/O resources, and I/O-bound in the opposite case; hence, a processor-bound process is mostly runnable, whereas an I/O-bound process is mostly blocked. As examples, crunching numbers is processor-bound, and accessing files is I/O-bound. Although an entire process might be characterized as either processor-bound or I/O-bound, a given process may be one or the other during different stages of its execution. Interactive desktop applications, such as browsers, tend to be I/O-bound.

A good process scheduler has to balance the needs of processor-bound and I/O-bound tasks, especially in an operating system such as Linux that thrives on so many hardware platforms: desktop machines, embedded devices, mobile devices, server clusters, supercomputers, and more.

Unix popularized classic preemptive scheduling, which other operating systems including VAX/VMS, Windows NT, and Linux later adopted. At the center of this scheduling model is a fixed timeslice, the amount of time (e.g., 50ms) that a task is allowed to hold a processor until preempted in favor of some other task. If a preempted process has not finished its work, the process must be rescheduled. This model is powerful in that it supports multitasking (concurrency) through processor time-sharing, even on the single-CPU machines of yesteryear.

The classic model typically includes multiple scheduling queues, one per process priority: Every process in a higher-priority queue gets scheduled before any process in a lower-priority queue. As an example, VAX/VMS uses 32 priority queues for scheduling.

CFS dispenses with fixed timeslices and explicit priorities. The amount of time for a given task on a processor is computed dynamically as the scheduling context changes over the system's lifetime. Here is a sketch of the motivating ideas and technical details:

Imagine a processor, P, which is idealized in that it can execute multiple tasks simultaneously. For example, tasks T1 and T2 can execute on P at the same time, with each receiving 50% of P's magical processing power. This idealization describes perfect multitasking, which CFS strives to achieve on actual as opposed to idealized processors. CFS is designed to approximate perfect multitasking.

The CFS scheduler has a target latency, which is the minimum amount of time—idealized to an infinitely small duration—required for every runnable task to get at least one turn on the processor. If such a duration could be infinitely small, then each runnable task would have had a turn on the processor during any given timespan, however small (e.g., 10ms, 5ns, etc.). Of course, an idealized infinitely small duration must be approximated in the real world, and the default approximation is 20ms. Each runnable task then gets a $1/N$ slice of the target latency, where N is the number of tasks. For example, if the target latency is 20ms and there are four contending tasks, then each task gets a timeslice of 5ms. By the way, if there is only a single task during a scheduling event, this lucky task gets the entire target latency as its slice. The fair in CFS comes to the fore in the $1/N$ slice given to each task contending for a processor.

The 1/N slice is, indeed, a timeslice—but not a fixed one because such a slice depends on N, the number of tasks currently contending for the processor. The system changes over time. Some processes terminate and new ones are spawned; runnable processes block and blocked processes become runnable. The value of N is dynamic and so, therefore, is the 1/N timeslice computed for each runnable task contending for a processor. The traditional nice value is used to weight the 1/N slice: a low-priority nice value means that only some fraction of the 1/N slice is given to a task, whereas a high-priority nice value means that a proportionately greater fraction of the 1/N slice is given to a task. In summary, nice values do not determine the slice, but only modify the 1/N slice that represents fairness among the contending tasks.

The operating system incurs overhead whenever a context switch occurs; that is, when one process is preempted in favor of another. To keep this overhead from becoming unduly large, there is a minimum amount of time (with a typical setting of 1ms to 4ms) that any scheduled process must run before being preempted. This minimum is known as the minimum granularity. If many tasks (e.g., 20) are contending for the processor, then the minimum granularity (assume 4ms) might be more than the 1/N slice (in this case, 1ms). If the minimum granularity turns out to be larger than the 1/N slice, the system is overloaded because there are too many tasks contending for the processor—and fairness goes out the window.

When does preemption occur? CFS tries to minimize context switches, given their overhead: time spent on a context switch is time unavailable for other tasks. Accordingly, once a task gets the processor, it runs for its entire weighted 1/N slice before being preempted in favor of some other task. Suppose task T1 has run for its weighted 1/N slice, and runnable task T2 currently has the lowest virtual runtime (vruntime) among the tasks contending for the processor. The vruntime records, in nanoseconds, how long a task has run on the processor. In this case, T1 would be preempted in favor of T2.

The scheduler tracks the vruntime for all tasks, runnable and blocked. The lower a task's vruntime, the more deserving the task is for time on the processor. CFS accordingly moves low-vruntime tasks towards the front of the scheduling line. Details are forthcoming because the line is implemented as a tree, not a list.

How often should the CFS scheduler reschedule? There is a simple way to determine the scheduling period. Suppose that the Target Latency (TL) is 20ms and the Minimum Granularity (MG) is 4ms:

TL / MG = (20 / 4) = 5 ## five or fewer tasks are ok

In this case, five or fewer tasks would allow each task a turn on the processor during the target latency. For example, if the task number is five, each runnable task has a 1/N slice of 4ms, which happens to equal the minimum granularity; if the task number is three, each task gets a 1/N slice of almost 7ms. In either case, the scheduler would reschedule in 20ms, the duration of the target latency.

Trouble occurs if the number of tasks (e.g., 10) exceeds TL / MG because now each task must get the minimum time of 4ms instead of the computed 1/*N* slice, which is 2ms. In this case, the scheduler would reschedule in 40ms:

(number of tasks) * MG = (10 * 4) = 40ms ## period = 40ms

Linux schedulers that predate CFS use heuristics to promote the fair treatment of interactive tasks with respect to scheduling. CFS takes a quite different approach by letting the vruntime facts speak mostly for themselves, which happens to support sleeper fairness. An interactive task, by its very nature, tends to sleep a lot in the sense that it awaits user inputs and so becomes I/O-bound; hence, such a task tends to have a relatively low vruntime, which tends to move the task towards the front of the scheduling line.

4.5 实验4 Linux调度策略

获取视频

1. 实验目的

了解Linux的进程/线程调度策略，通过终端指令观察实验结果，理解策略。

2. 实验方法

本实验内容的理论部分全部来自Linux下的sched帮助文档，相对比较枯燥。实验指导翻译了其中的概述部分，其余内容在实验前请读者自行阅读；对照实验内容观看视频，最后自行完成实验操作，观察其实验结果。

3. 实验指导

每个Linux线程都有一个与之关联的调度策略和静态优先级sched_priority，调度器依据这个策略和静态优先级进行调度决策。调度策略分为normal scheduling policy和real-time policy两种。

(1) normal scheduling policy：该种调度策略下有SCHED_OTHER、SCHED_IDLE及SCHED_BATCH三种方案，使用这种调度策略时静态优先级sched_priority始终为0。

(2) real-time policy：该种策略有SCHED_FIFO和SCHED_RR两种方案，其sched_priority的取值范围为[1, 99]，该数值和线程优先级成正比，因此real-time线程的优先级总比normal线程高。

4. 实验内容(观看视频后完成)

(1) 掌握ps指令，观察多线程tid和数量的方法；

(2) 理解NICE值在计算线程优先数中的作用；

(3) PR值计算公式：PR = −1−sched_priority，理解PR的取值范围与优先级之间的关系；

(4) 掌握chrt指令，观察不同进程调度策略的方法。

第5章 进程同步

获取视频

5.1 背景

在实验 3 中，我们实现了一个多线程的订票系统，两个订票线程并发运行，剩余票数记录在全局变量 ticketAmount 中，初始值为 2，根据多次执行该代码的实验结果来看，ticketAmount 的值不唯一，时而为 0、时而为 1。本章再以这个例子为背景，引出话题：进程同步。

假设一个订票系统有两个终端，分别运行进程 T1 和 T2。该系统公用存储区中的单元 B 存放某日某次航班机票的余量，售票代码如下：

```
process Ti(i = 1,2){
    int di;
    di = B;
    if (di >= 1){
        di = di -1 ;
        B = di;
    }
}
```

由于 T1 和 T2 是两个可同时运行的并发进程，也就意味着它们的运行是可以交替的，B 作为公用存储单元为两个进程所共享，所以可能出现以下的运行序列(B 初始值为 2)：

```
T1: d1 = B (B=2)
T2: d2 = B (B=2)
T2: d2 = d2 -1 = 2 - 1 = 1
T1: d1 = d1 -1 = 2 - 1 = 1
T1: B = d1 = 1;
T2: B = d2 = 1;
```

因为交替的过程是不可预知的，当然也可能出现第二种运行序列(B 初始值为 2)：

```
T1: d1 = B (B=2)
T1: d1 = d1 -1 = 2 - 1 = 1
T1: B = d1 = 1;
T2: d2 = B (B=1)
T2: d2 = d2 -1 = 1 - 1 = 0
T2: B = d2 = 0;
```

还可以写出第三种运行序列(B 初始值为 2)：

```
T1: d1 = B (B=2)
T1: d1 = d1 -1 = 2 - 1 = 1
T2: d2 = B (B=2)
```

```
T2: d2 = d2 -1 = 2 - 1 = 1
T1: B = d1 = 1;
T2: B = d2 = 1;
```

还有其他可能的运行序列，除了第二种之外的其他运行序列得到的结果都是不正确的：票卖出了 2 张，但余票仍然为 1。对于一个信息系统而言，这种错误是绝对不允许的，这种错误发生的原因在于，多个进程并发访问和操作同一数据且并发进程的运行序列不可预测，这种情况称为竞争条件(Race Condition)。

由于操作系统中的不同部分都要操作资源，因此开发多线程应用非常重要。在这类应用中，多个进程或线程很可能共享数据，并在不同的处理器上并发运行，这就要求这些进程或线程进行的任何操作不会互相干扰。因为现代操作系统中以线程为调度单位，进程同步和线程同步在算法理论上没有本质区别。本章内容不刻意区分二者的含义，有些地方还可能会混用两个词汇。

获取视频

5.2 进程的交互

并发进程之间的关系有两种：独立关系和交互关系。独立关系的并发进程是指进程之间不会因为运行序列的不同而对对方产生影响，它们不会共享公共区域和全局变量，如 Chrome 进程和 QQ 进程；交互关系的并发进程指进程之间有共享的公共区域和全局变量，它们对这些共享数据的读/写顺序对计算结果有很大的影响，前面提到的订票系统就是典型的交互进程。进程之间的关系见图 5-1。

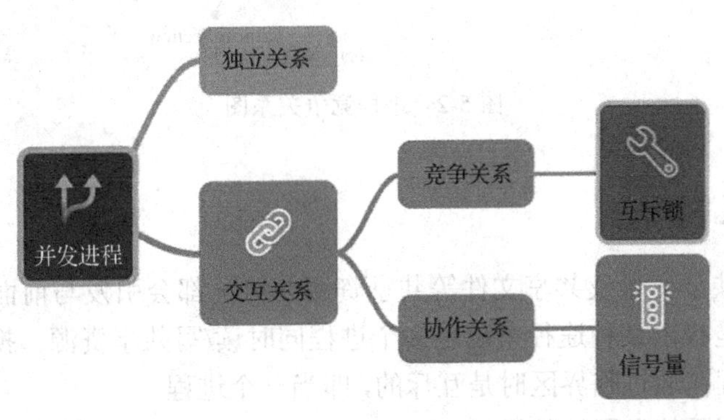

图 5-1　进程关系

很明显，我们主要要讨论的是交互关系的进程，并发进程间交互的方式主要有两种：竞争和协作。多个进程在运行过程中都需要某个资源时，它们便产生了竞争关系，它们可能竞争某一块内存空间，也可能竞争某一个 I/O 设备。当一方获取资源时，其他进程只能在该进程释放资源之后才能去访问该资源，这就是进程互斥，操作系统提供了互斥锁(Mutex Lock)完成这个功能。当两个进程运行时，进程 A 需要获取进程 B 此时运行到某一步的运行结果或者信息，才能进行自己的下一步工作，这时就得等待进程 B 与自己通信(发送某个消息或信号)，而后进程 A 再继续运行。这种进程之间相互等待对方发送消息或信号的协作关系就称为进程同步，实现进程同步的工具有很多，本书只介绍其中一种——信号量。

5.3 竞争关系

5.3.1 竞争

图 5-2(a)展示了一个完整的竞争关系，两辆车在各自的轨道上行驶，在中间的岔路口发生竞争，如果不加以干预，在某些情况下两辆车可能会撞车，这不是我们想看到的。一种解决竞争的方式是在有竞争的地方设置一个标记(可以是一面红旗)，当一辆车先到达岔路口时设置这个标记，另一辆车看到标记会自动停下等待，并隔一段时间再检测标记是否移除，先行的车辆通过路口后将标记移除，等待车辆会在下一次检测时发现标记已移除，于是它可以占用该路口同时它也要设置标记……[如图 5-2(b)所示]这个发生竞争的区域称为“临界区”(Critical Section)，英文直译为“关键的区域”[如图 5-2(c)所示]。

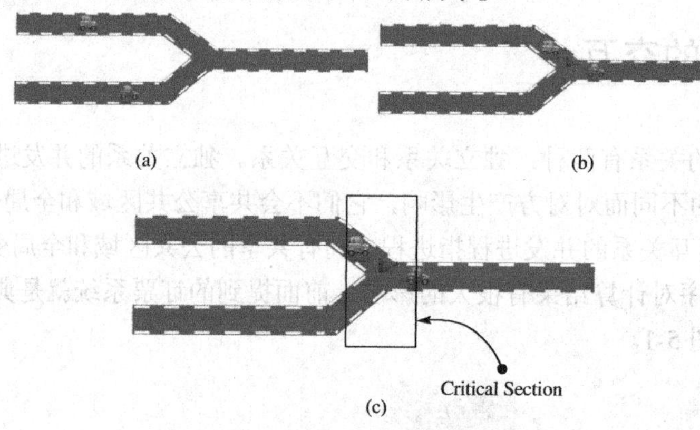

图 5-2　进程竞争关系图

5.3.2 临界区

实际上凡涉及共享变量及共享文件等共享资源的情况都会引发与前面类似的错误，要避免这种错误，关键是找出某种途径来阻止多个进程同时读/写共享资源。换言之，我们需要设计一个协议确保进程在通过临界区时是互斥的，即当一个进程在临界区时其他进程不能做同样的操作。

一种可行的协议如图 5-3 所示，该协议包括两部分：进入临界区前在 entry section 请求许可；离开临界区后在 exit section 归还许可。协议还附带三条规则。

(1) 互斥：任何两个进程不能同时处于其临界区，如果进程在 entry section 得不到许可，即说明有进程正在临界区中，该进程必须在此等待。

(2) 前进(Progress)：如果进程在 entry section 拿到了许可，即临界区是空闲的，那么进程就一定要前进至临界区，否则无法正常归还许可。

```
do {
        entry section
              critical section
        exit section

              remainder section
} while (true);
```

图 5-3　临界区管理协议

(3) 有限等待(Bounded Waiting)：进程不得无限期等待进入临界区，如果进程离开临界区，一定要在 exit section 归还许可，否则其他进程无法再进入临界区。

从抽象的角度看，我们所希望的进程使用临界区的行为如图 5-4 所示。在图 5-4 中，进程 A 在 T_1 时刻进入临界区。在 T_2 时刻进程 B 试图进入临界区，但是失败了，因为另一个进程已经在该临界区内，而一个时刻只允许一个进程在临界区内。因此，进程 B 阻塞，直到 T_3 时刻进程 A 离开临界区，随后进程 B 立即进入。最后，进程 B 在 T_4 时刻离开临界区，回到在临界区中没有进程的原始状态。

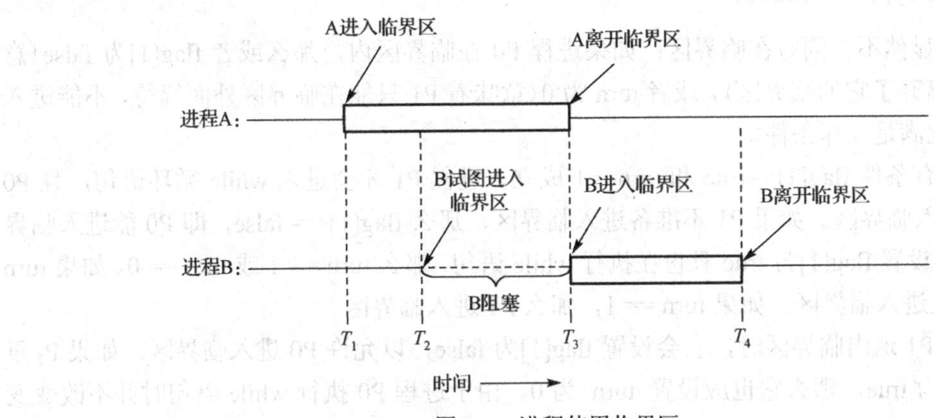

图 5-4　进程使用临界区

5.3.3　软件解决方案

下面介绍一个经典的基于软件的临界区问题解决方案，称为 Peterson 解决方案。算法使用两个控制变量 flag 与 turn，其中 flag[n]的值为真，表示 ID 号为 n 的进程希望进入该临界区，变量 turn 保存有权访问共享资源的进程 ID 号，while 循环表示 entry section，最后的 flag[n]=false 表示 exit section，完整算法代码如下：

```
//flag[] is boolean array; and turn is an integer
flag[0] = false;
flag[1] = false;
int turn;
P0:
flag[0] = true;
    turn = 1;
    while(flag[1] == true && turn == 1)
    {
        // busy wait
    }
    // critical section
    ...
    // end of critical section
    flag[0] = false;
P1:
flag[1] = true;
```

```
    turn = 0;
    while(flag[0] == true && turn == 0)
    {
        // busy wait
    }
    // critical section
    ...
    // end of critical section
    flag[1] = false;
```

P0 与 P1 显然不会同时在临界区：如果进程 P0 在临界区内，那么或者 flag[1]为 false(意味着 P1 已经离开了它的临界区)，或者 turn 为 0(意味着 P1 只能在临界区外面等待，不能进入临界区)，因此满足互斥条件。

另外，只有条件 flag[1]==true 和 turn==1 成立，进程 P1 才会进入 while 循环语句，且 P0 就能被阻止进入临界区。如果 P1 不准备进入临界区，那么 flag[1] == false，即 P0 能进入临界区。如果 P1 已设置 flag[1]为 true 且也在执行 while 语句，那么 turn == 1 或 turn == 0。如果 turn == 0，那么 P0 进入临界区。如果 turn == 1，那么 P1 进入临界区。

然而，当 P1 退出临界区时，它会设置 flag[1]为 false，以允许 P0 进入临界区。如果 P_1 重新设置 flag[1]为 true，那么它也应设置 turn 为 0。由于进程 P0 执行 while 语句时并不改变变量 turn 的值，所以 P0 会进入临界区(满足前进条件)，而且 P0 在 P1 进入临界区后最多一次就能进入(满足有限等待条件)。

如果并发进程数量多于 2，这个算法的复杂度会进一步提升，因此基于软件的解决方案(如 Peterson 解决方案)并不确保能在现代计算机体系结构上正确工作。在现代操作系统中是通过“锁”来保护临界区的，锁通常由硬件指令实现。

5.3.4 硬件解决方案

如果我们可以采用硬件锁来管理临界区，那么临界区的管理问题可以抽象成图 5-5 所示的模型。进入临界区前先通过 lock 获取锁，如果获取失败，则等待；离开临界区后通过 unlock 释放锁。

如何用硬件指令实现锁？在单处理器系统上，最常见的方式就是关闭尽可能多的可能对共享数据段进行读/写的指令中断。这样一来，就可以避免在临界区中暂停程序运行，或来自硬件的要求修改目标共享数据段的中断请求。但是关闭中断这个方案不适用于多处理器系统，多处理器的中断禁止很耗时，因为消息要传递给所有处理器。消息传递会使进程延迟进入临界区，降低系统效率。另外，如果系统时钟是通过中断来更新的，那么它也会受到影响。

```
Pi:
lock()
//critical section
unlock()
```

图 5-5　硬件锁管理临界区模型

许多现代操作系统提供特殊的硬件指令用于检测和修改文字的内容，这些指令都是原子(Atomic)指令，简称为“原语”。原子操作的特点是，其在执行过程中是不可被中断的，我们可以利用这个特点来实现 lock 操作，这里主要讲解使用 test_and_set 指令实现 lock 操作的过程。

首先，test_and_set 指令可以按以下代码定义，这一指令的执行是原子的，也就是说，如果两个 test_and_set 指令同时执行在不同的 CPU 上，那么它们可以按任意次序顺序执行，而不会给对方造成影响。

```
bool test_and_set(bool target){
    bool re = target;
    target = true;
    return re;
}
```

那么临界区管理中的 lock 和 unlock 实现见如下代码，available 表示临界区的可用状态，true 表示临界区可用，即未上锁，false 则表示相反的意思。

```
bool available = true;//unlocked
lock(){
  while(!test_and_set(&available))
        do nothing;
}

unlock(){
  available = true;
}
```

因为 test_and_set 是原子操作，所以 lock 中的测试及上锁过程是不会被中断的，也就避免了其他并发进程打断这个过程，满足了临界区管理的要求。本章的实验 5 将使用 Linux 提供的互斥锁对之前的订票系统实现临界区管理。

5.3.5 忙式等待

上述解决方案在本质上是这样的：当一个进程想进入临界区时，先检查是否允许其进入，若不允许，则该进程将原地等待[称为忙式等待(Busy Waiting)]，直到允许为止。其实，这种类型的互斥锁称为自旋锁(Spin Lock)，因为进程不停地“旋转”，以等待锁可用。在实际多道批处理系统中，即当多个进程共享同一个 CPU 时，这种连续循环显然是个问题，会浪费 CPU 时间。不过，自旋锁有一个优点：当进程在等待锁时，没有上下文切换(上下文切换可能需要相当长的时间)。因此，当使用锁的时间较短时，自旋锁还是有用的。

消除忙式等待的方法是阻塞需要等待的进程，也就是让它们从运行态切换到等待态，在 unlock 中除了解锁，还要多一个唤醒等待进程的操作，让阻塞的进程进入就绪队列等待调度。

5.4 协作关系

获取视频

5.4.1 信号量

1965 年荷兰学者 Dijkstra 提出了信号量(Semaphore)的概念。信号量是一种比互斥锁更强大的同步工具，它可以提供更高级的方法来同步并发进程。简单来说，信号量是一个整数变量，但是除初始化赋值外，它只能被两个原子操作——P 操作及 V 操作控制。P 操作也写作 wait 操作，V 操作也写作 signal 操作，两个操作的定义代码如下(s 表示信号量)。

```
P(s){
  while(s<=0)
     do nothing;
  s--;
}
V(s){
  s++;
}
```

在信号量 s 的值大于 0 时，P 操作会对其进行减 1 操作，否则会进入一个忙式等待，V 操作简单地将信号量加 1。信号量的值是一个整数，在定义时可对其进行初始化赋值，如果信号量的值只能是 0 或 1，则称这种信号量为“二值信号量”(Binary Semaphore)，如果信号量的值不受限制则称为“计数信号量”(Counting Semaphore)。

5.4.2 二值信号量

若信号量的初始值为 1 且为二值信号量，这种信号量可以用来实现互斥锁的功能。

```
semaphore mutex = 1;
process pi{
   P(mutex);
   critical section
   V(mutex);
}
```

上述使用了二值信号量 mutex 及 P、V 操作实现了临界区管理，P(mutex)是上锁过程，而 V(mutex)是解锁过程。假设有两个进程 P1 和 P2 并发运行这段代码，当 P1 先执行 P(mutex)，因为 mutex 值此时为 1 会跳过循环，mutex 的值被减 1 后为 0，P 操作结束，P1 顺利进入临界区；P2 再执行 P(mutex)时会因为 mutex 值为 0 进入忙式等待，从而实现了临界区的互斥；P1 离开临界区后执行 V(mutex)会将 mutex 值从 0 恢复到 1；P2 下次循环检测到 mutex 大于 0，于是将其减 1 后进入临界区，离开临界区后恢复 mutex 的值为 1。

本章的实验 5 会要求读者使用二值信号量替换之前的互斥锁来解决订票系统的临界区管理问题。

5.4.3 计数信号量

计数信号量的值域不受限制，初始值通常大于或等于 1，用于控制并发进程对共享资源的访问。考虑一个如图 5-6 所示的问题，三辆车通过一个二分岔的路口，分岔路口只允许一辆车通过，很显然在这个分岔处需要进行一些控制。在这个例子中，车是进程，分岔路是资源，进程数量大于资源数量，所以需要在分岔处进行同步，即当可用资源不足时，新申请资源的进程需要等待资源释放。

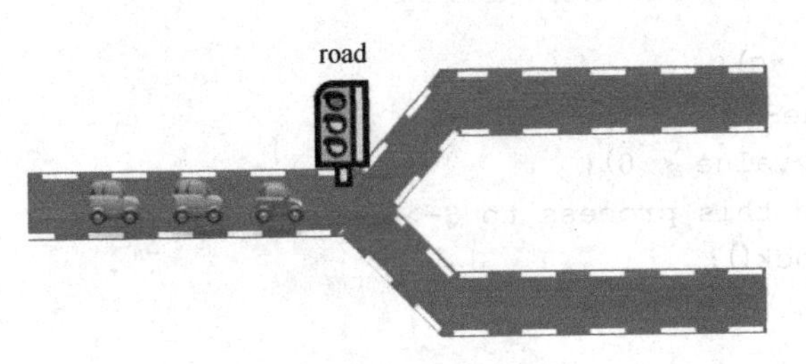

图 5-6　汽车过岔路问题

我们现在用信号量来完成二分岔控制管理，在分岔路口设置一盏信号灯 road，初始值为 2，表示两条分岔路都是空闲的，并发汽车进程记作 Car_i (i=1,2,3)，进程代码如下：

```
semaphore road = 2;
process Car_i{
   P(road);
   pass the fork in the road.
   V(road);
}
```

当第一辆车通过分岔路口时，road 值会被 P 操作减 1，类似地，第二辆车通过后，road 值被减为 0，当第三辆车驶来时，会因为 road 值为 0 而陷入忙式等待，直到前面任一辆车驶离后执行 V 操作将 road 值加 1。读者可以跟踪 road 值的变化，在三辆车全部驶离后，road 值会被恢复成初始值 2，这说明资源的总数没有变化，分配出去的全部被收回来了，这无疑是正确的。因此当你发现程序运行完，信号量的值没有恢复为初始值就说明有进程占用的资源未被释放。

实际上，这个问题和临界区管理问题没有本质区别，只是可用资源数量上有差异而已。因为临界区是互斥资源，我们可以将临界区视为可用数量为 1 的资源。

5.4.4　信号量的实现

5.4.1 节定义的信号量 P 操作也有与互斥锁类似的忙式等待现象，为了克服忙式等待，可以修改信号量 P、V 操作的定义：当一个进程执行 P 操作且发现信号量的值不为正时，它必须等待。然而，该进程不是忙式等待的而是阻塞的。阻塞操作将一个进程放到与信号量相关的等待队列中，并且将该进程状态切换成等待态。然后，将控制权转到 CPU 调度程序中，以便运行另一个进程。因等待信号量而阻塞的进程，在其他进程执行 V 操作后，应重新运行。进程的重新运行是通过 wakeup 操作进行的，它将进程从等待态切换为就绪态，被插入就绪队列中。

因为需要设计一个等待队列，我们用链表实现，于是将信号量数据类型设计如下：

```
typedef struct{
    int value;
    struct process *list;
} semaphore;
```

每个信号量都包括一个整型 value 和一个进程链表指针 list。当一个进程必须等待信号量时，就会被添加到进程链表中。V 操作从进程链表上取走一个进程，并加以唤醒。信号量 P、V 操作定义如下：

```
P(semaphore *S){
    S->value--;
    if  (s->value < 0){
        add this process to S->list;
        block();
    }
}
V(semaphore *S){
    S->value++;
    if  (s->value <= 0){
        remove a process p from S->list
        wakeup(p);
    }
}
```

block()用于挂起调用它的进程，wakeup(p)唤醒阻塞进程p。注意，这样实现的信号量的值可以是负数，而它的绝对值就是等待它的进程数。

5.4.5 死锁与饥饿

假设有一个系统，它有两个进程P0和P1，访问共享信号量S和Q，这两个信号量的初值均为1：

```
P0          P1
P(S);       P(Q);
P(Q);       P(S);
 ⋮           ⋮
V(S);       V(Q);
V(Q);       V(S)
```

假设P0执行P(S)，接着P1执行P(Q)，当P0执行P(Q)时，它必须等待，因为Q为0，直到P1执行V(Q)。同样地，P1执行P(S)时，它也必须等待，因为S为0，直到P0执行V(S)。由于这两个V操作都无法执行，所以P0和P1会无限等待下去，这样的状态就是死锁(Deadlock)。

与死锁相关的另一种说法是饥饿(Starvation)，饥饿与死锁的区别是，饥饿是指长时间等待，可能有结束的时候，而死锁是无限等待。

5.5 经典同步问题

获取视频

5.5.1 最简单的同步问题

同步问题的实质是将异步的并发进程按照某种顺序运行，这些异步的并发进程都是交互关系，因此找到它们的交互点至关重要，可充分利用P操作的等待特点来调节进程的运行速度。通常初始值0的信号量可以让进程直接进入等待状态直到另一个进程唤醒它。

举个例子，有两个并发进程：司机和售票员；司机进程的动作是，启动车辆、正常行车及到站停车；售票员的动作是，关车门、售票及开车门。同步要求是，司机要等售票员关车门才能开车；售票员要等司机停车才能开车门。两个进程的同步代码如下：

```
semaphore C=0, D=0
Driver:                                         Conductor:
  P(D)                                            关车门
  启动车辆                                         V(D)
  正常行车                                         售票
  到站停车                                         P(C)
  V(C)                                            开车门
```

对于司机而言，启动车辆受到售票员关车门的影响，两个动作有先后顺序，所以司机在启动车辆先执行 P(D)，因为 D 的初始值为 0，所以司机会立即陷入等待，直到售票员关上车门，执行 V(D)；类似地，售票员在开车门前执行 P(C)，陷入等待，直到司机停车后执行 V(C)。

这个同步问题和之前的临界区管理问题最大的区别在于，该同步问题中的同一个信号量的 P、V 操作在两个不同的进程中，而临界区管理的信号量 P、V 操作在相同的进程中。在复杂问题中，这两种信号量操作大多会同时出现，读者在研究新问题时，只要抓住问题的本质就不至于找不到头绪。

5.5.2 生产者-消费者问题

生产者-消费者问题是一个非常经典的同步问题，本质上它是一个有界的缓冲问题（Bounded-Buffer Problem），生产者(P)与消费者(C)公用一个缓冲区，生产者不能往“满”的缓冲区中放产品，消费者不能从“空”的缓冲区中取产品。

首先从最简单的单缓冲学习，生产者及消费者公用一个只能容纳一个产品的缓冲区，解决方案如下：

```
semaphore empty = 1;
semaphore full = 0;
Producer{
  while (true){
    make a product;
    P(empty);
    put the product into buffer;
    V(full);
  }
}
Consumer{
  while (true){
     P(full);
     pick product from buffer;
     V(empty);
     consume the product;
  }
}
```

如果将缓冲区扩展成容纳 *k* 个产品(如图 5-7 所示)，则需要以下数据结构：

```
item B[k];
semaphore mutex = 1;
semaphore empty = k;
semaphore full  = 0
int in=0, out=0;
```

图 5-7　有界缓冲区问题

B[k]是用来表示缓冲区的数组，共有 k 个空间，in 和 out 分别给生产者及消费者指示 B 数组下标，信号量 mutex 提供缓冲区访问的互斥要求，并被初始化为 1。信号量 empty 和 full 分别用于表示空的和满的缓冲区数量。信号量 empty 被初始化为 k，而信号量 full 被初始化为 0。

生产者进程代码如下：

```
Process producer_i{
     make a product;
     P(empty);
     P(mutex);
     B[in] = product;
     in = (in+1) % k;
     V(mutex)
     V(full);
}
```

消费者进程代码如下：

```
Process consumer_i{
     P(full);
     P(mutex)
     product = B[out];
     out = (out+1) % k;
     V(mutex)
     V(empty);
     consume a product;
}
```

5.5.3　苹果桔子问题

问题描述：如图 5-8 所示，桌上有一只盘子，每次只能放入一个水果；爸爸专向盘子中放苹果，妈妈专向盘子中放桔子；儿子专等吃盘子中的桔子，女儿专等吃盘子中的苹果。

先简单分析一下：这是一个单缓冲问题，但是产品的种类有两种，有两个生产者及两个消费者。

解决方案代码如下：

图 5-8　苹果桔子问题

```
semaphore sp     = 1;   /* 盘子里允许放一个水果*/
semaphore sg1    = 0;   /* 盘子里没有桔子 */
semaphore sg2    = 0;   /* 盘子里没有苹果*/
Process father{
    削一个苹果;
    P(sp);
    把苹果放入盘子;
    V(sg2);
}
Process mother{
    剥一个桔子;
    P(sp);
    把桔子放入盘子;
    V(sg1);
}
Process daughter{
    P(sg2);
    从盘子中取苹果;
    V(sp);
    吃苹果;
}
Process son{
    P(sg1);
    从盘子中取桔子;
    V(sp);
    吃桔子;
}
```

5.5.4　哲学家进餐问题

如图 5-9 所示，有 5 个哲学家，他们的生活包括思考和吃面两部分，哲学家们坐在一张在圆桌前，上面有 5 碗意大利面和 5 把叉子。当哲学家思考时会放下手中的叉子，当他想吃面时必须要拿到左、右两把叉子才可以吃，当叉子在邻座哲学家手上时，必须等待他放下才可以取。

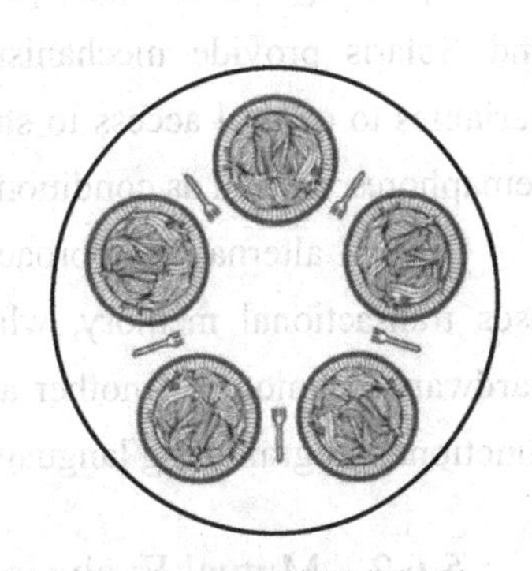

图 5-9　哲学家进餐问题

哲学家进餐问题(Dining-Philosophers Problem)是一个经典的同步问题，这不是因为其本身的重要性，而是因为它是大量并发控制问题的例子。这个例子最重要的情况是 5 个哲学家同时饥饿并拿起左边的叉子，这将导致所有哲学家都在等待邻座释放叉子从而进入“死锁”状态。

这个死锁问题有多种可能的解决方案：

(1)最多允许4个哲学家同时坐在桌子上；

(2)只有当一个哲学家的两把叉子都可用时，他才能拿起它们，否则他一把也不拿；

(3)使用非对称解决方案，即单号的哲学家先拿左边的叉子，接着再拿右边的叉子；而双号的哲学家先拿右边的叉子，接着再拿左边的叉子。

死锁问题在并发进程的同步控制中时常发生，死锁现象出现的原因及解决方案将在下一章阐述。

5.6 Reading Materials

获取视频

5.6.1 Overview

Given a collection of cooperating sequential processes that share data, mutual exclusion must be provided to ensure that a critical section of code is used by only one process or thread at a time. Typically, computer hardware provides several operations that ensure mutual exclusion. However, such hardware based solutions are too complicated for most developers to use. Mutex locks and semaphores overcome this obstacle. Both tools can be used to solve various synchronization problems and can be implemented efficiently, especially if hardware support for atomic operations is available.

Various synchronization problems(such as the bounded-buffer problem, the readers-writers problem, and the dining-philosophers problem) are important mainly because they are examples of a large class of concurrency-control problems. These problems are used to test nearly every newly proposed synchronization scheme.

The operating system must provide the means to guard against timing errors, and several language constructs have been proposed to deal with these problems. Monitors provide a synchronization mechanism for sharing abstract data types. A condition variable provides a method by which a monitor function can block its execution until it is signaled to continue.

Operating systems also provide support for synchronization. For example, Windows, Linux, and Solaris provide mechanisms such as semaphores, mutex locks, spinlocks, and condition variables to control access to shared data. The Pthreads API provides support for mutex locks and semaphores, as well as condition variables.

Several alternative approaches focus on synchronization for multicore systems. One approach uses transactional memory, which may address synchronization issues using either software or hardware techniques. Another approach uses the compiler extensions offered by OpenMP. Finally, functional programming languages address synchronization issues by disallowing mutability.

5.6.2 Mutual Exclusion

During concurrent execution of processes, processes need to enter the critical section (or the section of the program shared across processes) at times for execution. It might so happen that because of the execution of multiple processes at once, the values stored in the critical section become inconsistent. In other words, the values depend on the sequence of execution of instructions - also known as a race condition. The primary task of process synchronization is to get rid of race conditions while executing the critical section.

This is primarily achieved through mutual exclusion.

Mutual exclusion is a property of process synchronization which states that "no two processes can exist in the critical section at any given point of time". The term was first coined by Djikstra. Any process synchronization technique being used must satisfy the property of mutual exclusion, without which it would not be possible to get rid of a race condition.

5.6.3 Critical Section

When more than one processes access a same code segment that segment is known as critical section. Critical section contains shared variables or resources which are needed to be synchronized to maintain consistency of data variable.

In simple terms a critical section is group of instructions/statements or region of code that need to be executed atomically (read this post for atomicity), such as accessing a resource (file, input or output port, global data, etc.).

In concurrent programming, if one thread tries to change the value of shared data at the same time as another thread tries to read the value (i.e. data race across threads), the result is unpredictable.

The access to such shared variable (shared memory, shared files, shared port, etc…) to be synchronized. Few programming languages have built-in support for synchronization.

It is critical to understand the importance of race condition while writing kernel mode programming (a device driver, kernel thread, etc.). since the programmer can directly access and modifying kernel data structures.

5.6.4 Mutex VS Semaphore

What are the differences between Mutex vs Semaphore? When to use mutex and when to use semaphore?

Concrete understanding of Operating System concepts is required to design/develop smart applications. Our objective is to educate the reader on these concepts and learn from other expert geeks.

As per operating system terminology, mutex and semaphore are kernel resources that provide synchronization services (also called as synchronization primitives). Why do we need such synchronization primitives? Won't be only one sufficient? To answer these questions, we need to understand few keywords. Please read the posts on atomicity and critical section. We will illustrate with examples to understand these concepts well, rather than following usual OS textual description.

The producer-consumer problem:

Consider the standard producer-consumer problem. Assume, we have a buffer of 4096 byte length. A producer thread collects the data and writes it to the buffer. A consumer thread processes the collected data from the buffer. Objective is, both the threads should not run at the same time.

Using Mutex:

A mutex provides mutual exclusion, either producer or consumer can have the key (mutex) and proceed with their work. As long as the buffer is filled by producer, the consumer needs to wait, and vice versa.

At any point of time, only one thread can work with the entire buffer. The concept can be generalized using semaphore.

Using Semaphore:

A semaphore is a generalized mutex. In lieu of single buffer, we can split the 4 KB buffer into four 1 KB buffers (identical resources). A semaphore can be associated with these four buffers. The consumer and producer can work on different buffers at the same time.

Misconception:

There is an ambiguity between binary semaphore and mutex. We might have come across that a mutex is binary semaphore. But they are not! The purpose of mutex and semaphore are different. May be, due to similarity in their implementation a mutex would be referred as binary semaphore.

Strictly speaking, a mutex is locking mechanism used to synchronize access to a resource. Only one task (can be a thread or process based on OS abstraction) can acquire the mutex. It means there is ownership associated with mutex, and only the owner can release the lock (mutex).

Semaphore is signaling mechanism ("I am done, you can carry on" kind of signal). For example, if you are listening songs (assume it as one task) on your mobile and at the same time your friend calls you, an interrupt is triggered upon which an interrupt service routine (ISR) signals the call processing task to wake up.

5.7 实验5 并发线程互斥同步

获取视频

1. 实验目的

掌握Pthread库的互斥锁的使用方法并用之解决临界区管理问题，掌握POSIX信号量的使用方法并用之解决互斥同步问题。

2. 实验方法

(1)阅读互斥锁及信号量相关API的手册，掌握调用方法；

(2)掌握P、V操作对应的POSIX信号量函数的调用方法；

(3)解决经典的生产者-消费者问题；

(4)解决其他同步问题。

3. 实验内容

(1) 接实验 3 中的多线程订票系统，找出代码的临界区，用 Pthread 互斥锁对临界区进行管理，阅读并运行下面的代码。

```
#include <stdio.h>
#include <pthread.h>
#include <unistd.h>
int ticketAmount = 2; //Global Variable
pthread_mutex_t lock = PTHREAD_MUTEX_INITIALIZER; //Global lock
void* ticketAgent(void* arg){
    pthread_mutex_lock(&lock);
    int t = ticketAmount;
    if(t > 0)
    {
        printf("One ticket sold!\n");
        t--;
    }else{
        printf("Ticket sold out!!\n");
    }
    ticketAmount = t;
    pthread_mutex_unlock(&lock);
    pthread_exit(0);
}
int main(int argc, char const *argv[])
{
    pthread_t ticketAgent_tid[2];
    for(int i = 0; i < 2; ++i)
    {
        pthread_create(ticketAgent_tid+i, NULL, ticketAgent,NULL);
    }
    for(int i = 0; i < 2; ++i)
    {
        pthread_join(ticketAgent_tid[i],NULL);
    }
    printf("The left ticket is %d\n", ticketAmount);
    return 0;
}
```

代码说明如下。

- pthread_mutex_t lock = PTHREAD_MUTEX_INITIALIZER;——创建一个互斥锁。
- pthread_mutex_lock (&lock);——上锁。
- pthread_mutex_unlock (&lock);——开锁。

(2) 使用 POSIX 信号量解决单缓冲问题。生产者和消费者共享一个缓冲区，代码如下：

```
#include <stdio.h>
#include <pthread.h>
#include <semaphore.h>
```

```
sem_t empty;
sem_t full;
void* producerThd(void* arg){
    for(int i=0; i<10; i++){
        printf("**Producing one item.**\n");
        sem_wait(&empty);
        printf("**PUTTING item to warehouse.**\n");
        sem_post(&full);
    }
    pthread_exit(0);
}
void* consumerThd(void* arg){
    for(int i=0; i<10; i++){
        sem_wait(&full);
        printf("##GETTING item from warehouse.##\n");
        sem_post(&empty);
        printf("##Consuming the item.##\n");
    }
    pthread_exit(0);
}
int main(int argc, char *argv[]){
    pthread_t producer_tid, consumer_tid;
    sem_init(&empty, 0, 1);
    sem_init(&full, 0, 0);
    pthread_create(&producer_tid, NULL, producerThd, NULL);
    pthread_create(&consumer_tid, NULL, consumerThd, NULL);
    pthread_join(producer_tid, NULL);
    pthread_join(consumer_tid, NULL);
    sem_destroy(&empty);
    sem_destroy(&full);
}
```

获取视频

代码说明如下。

- 要使用信号量，应先包含头文件<semaphore.h>。
- sem_t：信号量的数据类型。
- int sem_init(sem_t *sem, int pshared, unsigned int val);——该函数第一个参数为信号量指针，第二个参数为信号量类型（一般设置为 0），第三个参数为信号量初始值。当第二个参数 pshared 为 0 时，该进程内所有线程可用；不为 0 时，不同进程可用。
- int sem_wait(sem_t *sem); ——该函数申请一个信号量，当前无可用信号量时等待，有可用信号量时占用一个信号量，将信号量的值减 1。
- int sem_post(sem_t *sem);——该函数释放一个信号量，将信号量的值加 1。
- int sem_destory(sem_t *sem);——该函数销毁信号量。

(3)有界缓冲问题。当生产者和消费者共享多个缓冲区时，设 Bank[10]为仓库，有 10 个位置放置商品，元素为 0 表示无商品，为 1 表示有商品，除要用信号量同步外，还要加入互斥操作，保护临界区，代码如下：

```
#include <stdio.h>
#include <pthread.h>
#include <semaphore.h>
#include <unistd.h>
void printBank();
sem_t empty;
sem_t full;
int Bank[10]={0};
int in=0,out=0;
pthread_mutex_t lock = PTHREAD_MUTEX_INITIALIZER;
void* producerThd(void* arg){
    for(int i=0; i<20; i++){
        sem_wait(&empty);
        pthread_mutex_lock(&lock); //临界区开始
        Bank[in] = 1;
        in = (in+1)%10;
        printBank();
        sleep(0.1);
        pthread_mutex_unlock(&lock);//临界区结束
        sem_post(&full);
    }
    pthread_exit(0);
}
void* consumerThd(void* arg){
    for(int i=0; i<20; i++){
        sem_wait(&full);
        pthread_mutex_lock(&lock); //临界区开始
        Bank[out] = 0;
        out = (out+1)%10;
        printBank();
        sleep(1);
        pthread_mutex_unlock(&lock);//临界区结束
        sem_post(&empty);
    }
    pthread_exit(0);
}
/*该函数用以输出缓冲区的全部数值*/
void printBank(){
    printf("Bank:");
    for(int i=0; i<10; i++){
        printf("[%d]",Bank[i]);
        if(i==9) putchar('\n');
    }
}
int main(int argc, char *argv[]){
    pthread_t producer_tid, consumer_tid;
    sem_init(&empty, 0, 10);
```

```
    sem_init(&full, 0, 0);
    pthread_create(&producer_tid, NULL, producerThd, NULL);
    pthread_create(&consumer_tid, NULL, consumerThd, NULL);
    pthread_join(producer_tid, NULL);
    pthread_join(consumer_tid, NULL);
    sem_destroy(&empty);
    sem_destroy(&full);
}
```

我们用 sleep()来调节生产和消费的速度，让生产比消费快一点，这样可以更好地观察运行结果。

(4) 苹果桔子问题：有一个能放 N(这里 N 设为 3)个水果的盘子，采用以下三个信号量。

- sem_t empty：信号量 empty 控制盘子可放的水果数，初始值为 3，因为开始时盘子为空，可放水果数为 3。
- sem_t apple：信号量 apple 控制儿子可吃的苹果数，初始值为 0，因为开始时盘子里没有苹果。
- sem_t orange：信号量 orange 控制女儿可吃的桔子数，初始值为 0，因为开始时盘子里没有桔子。
- 互斥量 work_mutex 用于保证输出能够保持一致。

代码如下：

```
#include <stdio.h>
#include <pthread.h>
#include <semaphore.h>
#include <unistd.h>
sem_t empty;                //控制盘子可放的水果数
sem_t apple;                //控制儿子可吃的苹果数
sem_t orange;               //控制女儿可吃的桔子数
pthread_mutex_t work_mutex=PTHREAD_MUTEX_INITIALIZER; //声明互斥量work_mutex
int fruitCount = 0;         //盘子里水果的数量
void *procf(void *arg)      //father 线程
{
   while(1){
      sem_wait(&empty);                         //占用一个盘子空间,可放水果数减 1
      pthread_mutex_lock(&work_mutex);          //加锁
      printf("爸爸放入一个苹果!(盘子当前水果总数: %d)\n", ++fruitCount);
      sem_post(&apple);         //释放一个 apple 信号，可吃苹果数加 1
      pthread_mutex_unlock(&work_mutex);        //解锁
      sleep(0.1);
   }
}
void *procm(void *arg)                          //mother 线程
{
  while(1){
     sem_wait(&empty);
     pthread_mutex_lock(&work_mutex);           //加锁
```

```
        printf("妈妈放入一个橙子!(盘子当前水果总数：%d)\n", ++fruitCount);
        sem_post(&orange);
        pthread_mutex_unlock(&work_mutex);       //解锁
        sleep(0.2);
    }
}
void *procs(void *arg)                           //son线程
{
  while(1){
        sem_wait(&apple);                        //占用一个苹果信号量，可吃苹果数减1
        pthread_mutex_lock(&work_mutex);         //加锁
        printf("儿子吃了一个苹果!(盘子当前水果总数：%d)\n", --fruitCount);
        sem_post(&empty);        //吃了一个苹果，释放一个盘子空间，可放水果数加1
        pthread_mutex_unlock(&work_mutex);       //解锁
        sleep(0.2);
    }
}
void *procd(void *arg)                           //daughter线程
{
  while(1){
        sem_wait(&orange);
        pthread_mutex_lock(&work_mutex);         //加锁
        printf("女儿吃了一个桔子!(盘子当前水果总数：%d)\n", --fruitCount);
        sem_post(&empty);
        pthread_mutex_unlock(&work_mutex);       //解锁
        sleep(0.1);
    }
}
int main()
{
    pthread_t father;                            //定义线程
    pthread_t mother;
    pthread_t son;
    pthread_t daughter;
    sem_init(&empty, 0, 3);                      //信号量初始化
    sem_init(&apple, 0, 0);
    sem_init(&orange, 0, 0);
    pthread_create(&father,NULL,procf,NULL);     //创建线程
    pthread_create(&mother,NULL,procm,NULL);
    pthread_create(&daughter,NULL,procd,NULL);
    pthread_create(&son,NULL,procs,NULL);
    sleep(1);
    sem_destroy(&empty);
    sem_destroy(&apple);
    sem_destroy(&orange);
    return 0;
}
```

第6章 死　锁

获取视频

6.1 定义

在多道程序设计环境下，一些进程可能会竞争有限数量的资源，当进程请求的资源不可用时，会进入等待态，有些时候，这些等待的进程永远无法被唤醒，因为它们等待的资源也被处于等待态的进程占用，这种情况就称为“死锁”。

计算机系统的资源可以分成多种类型，每种类型有一定数量的实例。资源类型包括 CPU 周期、文件及 I/O 设备(如打印机)等。进程在使用资源前应申请资源，在使用资源后应释放资源。一个进程可能要申请许多资源，以便完成指定的任务。显然，申请的资源数量不能超过系统所有资源的总和。换言之，如果系统中只有两台打印机，那么进程就不能申请三台打印机。当一组进程内的一个进程在等待一个事件，而这一事件只能由这组进程内的另一个进程发起，那么这组进程就处于死锁状态。这里的事件主要是指资源的获取和释放。

图 6-1 演示了一个例子，进程 A 占用了扫描仪，正在申请打印机，进程 B 占用了打印机，正在申请扫描仪，两个进程都处于无限等待状态，这就是死锁，也可以理解为循环等待资源。

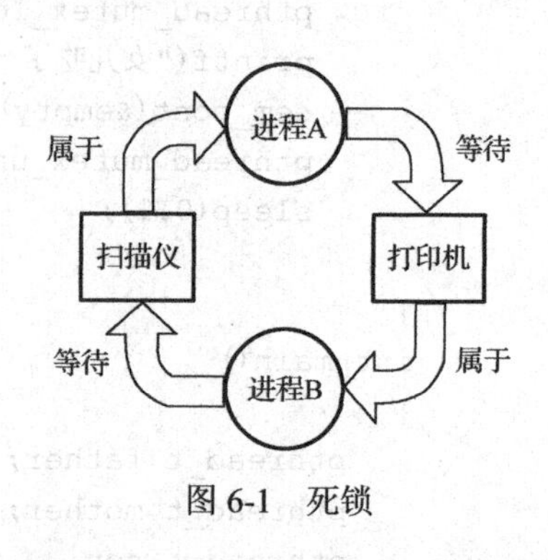

图 6-1　死锁

再假设一个系统中有三台 CD 刻录机。假设有三个进程，每个进程都占用了一台 CD 刻录机。如果每个进程现在需要另一台刻录机，那么这三个进程也会处于死锁状态。每个进程都在等待事件“CD 刻录机被释放”，这仅可能由一个等待进程来完成。这个例子说明了涉及同一种资源类型的死锁。

一组进程发生了死锁，对于计算机系统而言是不太好的，它们可能会拖累整个系统，在实际使用体验中，会出现鼠标、键盘长时间没有响应，界面切换卡顿及应用程序长时间没有响应等现象。死锁现象不是一个好现象，本章的内容就要讨论操作系统作为计算机的“管家”是如何对待死锁的。

6.2 死锁特征

如果在一个系统中，以下四个条件同时成立，那么就一定会引起死锁。这四个条件称为死锁的必要条件。

(1) 互斥(Mutual Exclusion)：至少有一个资源必须处于非共享模式(临界资源)，即一次只有一个进程可使用该资源。如果另一个进程申请该资源，那么申请进程应等到该资源被释放为止。

(2) 占用并等待(Hold and Wait)：一个进程应占用至少一个资源，并等待另一个被其他进程占用的资源。

(3) 非抢占(No Preemption)：资源不能被抢占，即资源只能被进程在完成任务后自愿释放。

(4) 循环等待(Circular Wait)：有一组等待进程$\{P_0, P_1, \cdots, P_n\}$，$P_0$等待的资源被$P_1$占用，$P_1$等待的资源被$P_2$占用，……，$P_n$等待的资源被$P_0$占用。

一般来说，操作系统设计人员对待死锁的态度分成三类。

(1) 极左派：不允许死锁出现，必须预防或避免系统进入死锁状态。因为四个必要条件要同时成立，死锁才会发生，所以只要破坏任一条件就可以防止(Prevent)死锁；或根据进程申请资源和使用资源的信息，依照算法进行推演后决定是否分配资源，从而达到避免(Avoid)死锁的目的。

(2) 中间派：允许死锁出现，但确保系统可以检测(Detect)到死锁并加以恢复。系统提供一种方法来检测系统状态以确定死锁是否发生，并提供另一种方法来从死锁中恢复(Recover)。

(3) 极右派：也称鸵鸟派，忽视死锁的存在，假装它不会发生。在这种情况下，未被发现的死锁会导致系统性能下降，因为资源被不能运行的进程占用，而越来越多的进程也会因申请资源而进入死锁。最后，整个系统会停止工作，只能人工重新启动。

第三种解决方案虽然看起来似乎不是一个解决死锁问题的可行方法，但是它为大多数操作系统所采用，包括 Linux 和 Windows。原因是，代价是一个重要的考虑因素，忽略死锁要比其他方法代价更小。对于许多系统，死锁很少发生(如一年一次)，因此，与使用频繁并且开销昂贵的死锁防止、死锁避免、死锁检测及恢复相比，这种方法开销更小。因此，应用程序开发人员需要自己编写程序，以处理死锁。

接下来，简要阐述三种死锁处理方法。有些研究人员认为，这些基本方法不能单独用于处理操作系统的所有资源分配问题。然而，可以将这些基本方法组合起来，为每种系统资源选择最佳方法。

6.3 资源分配图

通过资源分配图(Resource Allocation Graph, RAG)可以更精确地描述死锁。该图包括一个节点集合 V 和一个边集合 E，节点集合可分成两种类型：进程$T=\{T_1, T_2, \ldots, T_n\}$和资源类型$R=\{R_1, R_2, \ldots, R_n\}$。

从进程 T 到资源类型 R 的有向边记为 T→R，称为申请边(Request Edge)，它表示进程 T 已经申请了资源类型 R 的一个实例，并且正在等待这个资源。从资源类型 R 到进程 T 的有向边记为 R→T，称为分配边(Assignment Edge)，它表示资源类型 R 的一个实例已经分配给了进程 T。

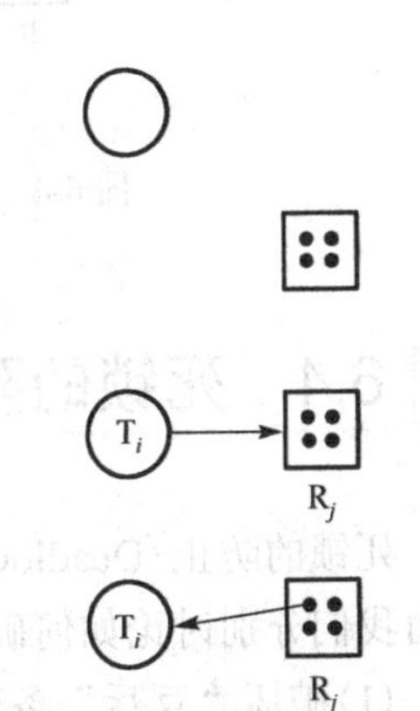

图 6-2 RAG 图例

如图 6-2 所示，用圆表示进程 T，用矩形表示资源类型 R。由于资源类型 R 可能有多个实例，所以矩形内点的数量表示实例数量。注意申请边只指向矩形 R，而分配边应指定矩形内的某个点。当进程 T 申请资源类型 R 的一个实例时，就在资源分配图中

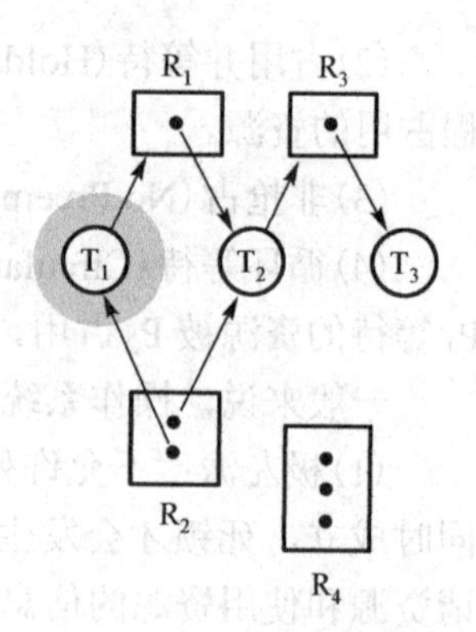

图 6-3　无环无死锁图

加入一条申请边。当进程不再需要访问资源时，它就会释放资源，删除分配边。

图 6-3 所示的资源分配图表示了如下情况：

$T = \{ T_1, T_2, T_3 \}$

$R = \{ R_1, R_2, R_3, R_4 \}$

$E = \{T_1 \rightarrow R_1, T_2 \rightarrow R_3, R_1 \rightarrow T_2, R_2 \rightarrow T_2, R_2 \rightarrow T_1, R_3 \rightarrow T_3\}$

资源实例：R_1 有一个实例，R_2 有两个实例，R_3 有一个实例，R_4 有三个实例。

这个资源分配状态中没有死锁，T_3 运行完毕后会释放 R_3，$T_2 \rightarrow R_3$ 申请边就可以消除，最终 T_2 释放 R_1，R_1 接着被分配给 T_1。可以证明，如果资源分配图中没有环，那么系统就没有死锁。如果资源分配图中有环，那么可能存在死锁。图 6-4 是一个有环有死锁的例子，$R_2 \rightarrow T_1 \rightarrow R_1 \rightarrow T_2 \rightarrow R_3 \rightarrow T_3 \rightarrow R_2$ 形成了一个环路，T_2 和 T_3 陷入无限等待。图 6-5 是一个有环无死锁的例子，$T_1 \rightarrow R_1 \rightarrow T_3 \rightarrow R_2 \rightarrow T_1$ 形成了一个环路，但是等 T_4 运行结束释放 R_2 后，T_1 的等待态就会消除，因此不会产生死锁。

所以，如果资源分配图中没有环路，那就必定没有死锁；如果资源分配图中出现了环路且每个资源类型刚好只有一个实例，那么意味着出现死锁；如果资源分配图中有环路且每类资源的实例数量大于1，那么就有可能出现死锁。

总而言之，如果资源分配图中没有环，那么系统就不会处于死锁状态。如果资源分配图中有环，那么系统可能会也可能不会处于死锁状态。在处理死锁问题时，这点是很重要的。

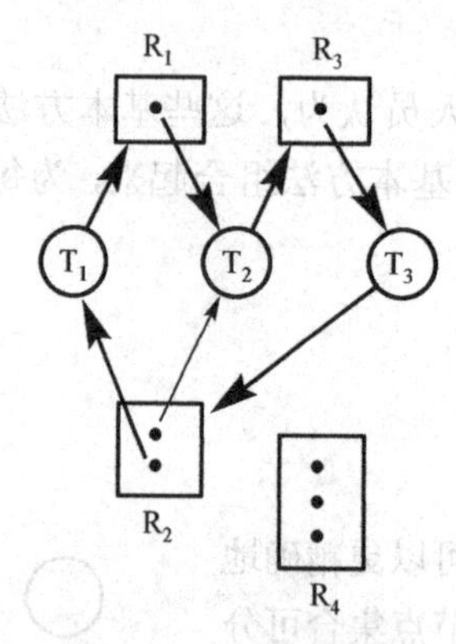

图 6-4　有环有死锁图

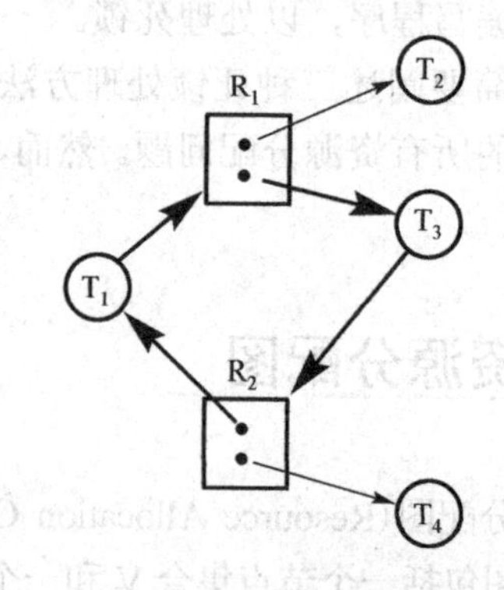

图 6-5　有环无死锁图

6.4　死锁的防止

死锁的防止（Deadlock Prevention）通过破坏四个必要条件之一达到防止死锁发生的目的，下面我们分别讨论如何破坏这些条件。

（1）破坏“互斥”条件。用反证法，假设该条件不成立，也就意味着所有的资源都是共享资源，但有些资源必须互斥使用，如打印机，因此该条件必须成立。

（2）破坏“占用并等待”条件。假设条件不成立，则应保证：当每个进程申请一个资源时，它不能占用其他资源。可以在每个进程运行前申请并获得所有资源，但是进程在整个运行过程中会占用过多资源导致资源的利用率下降，同时也有可能造成其他进程的“饥饿”现象。

(3)破坏“非抢占”条件。为确保这一条件不成立，可以规定：一个已持有若干资源的进程去申请另一个不能立即被分配的资源，那么它现有的资源可以被其他进程抢占。这个协议通常用于状态可以保存且恢复的资源，如CPU寄存器和内存。它一般不适用于其他资源，如互斥锁和信号量。

(4)破坏“循环等待”条件。确保这个条件不成立的一种方法如下：对所有资源类型进行完全排序，而且要求每个进程按递增顺序申请资源。以哲学家进餐问题为例，本章让他们用筷子吃米饭(如图6-6所示)。圆形表示哲学家，方形表示筷子，筷子的编号为从0到4，系统要求申请筷子时必须按照从低编号向高编号的顺序，也就是0号哲学家必须先拿0号筷子再拿1号筷子、1号哲学家必须先拿1号筷子再拿2号筷子、……、4号哲学家必须先拿0号筷子再拿4号筷子。一种极端的情况是，4号哲学家先去申请0号筷子，未成功就不会占用4号筷子，而3号哲学家就可以顺利地同时拿到一双筷子，破坏了循环等待条件。这种方法需要程序员按照顺序编写代码，提高了代码编写的复杂度，而且稍有不慎也会造成死锁。

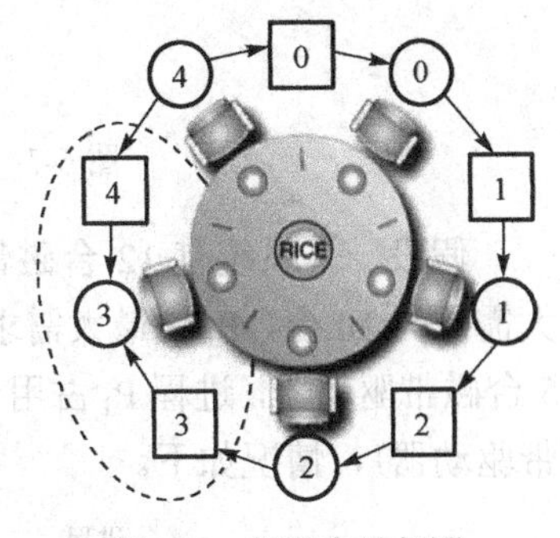

图6-6 哲学家进餐图

总体看来，死锁防止方案对资源的申请限制很高，对程序员编程技巧的要求也很高，而且会带来低下的设备利用率和系统吞吐量，因此缺乏实践意义。

6.5 死锁的避免

如果不去刻意破坏死锁的四个必要条件，那么可以在进程申请资源时使用算法进行推演，计算是否可以将资源分配给该进程从而避免死锁的发生。为了达到这个目标，系统需要事先知道进程在运行过程中所需要的资源种类及最大需求量，系统会通过一个算法确保系统不会进入死锁状态，这个算法就是死锁避免(Deadlock Avoidance)算法。

6.5.1 安全状态

当一个进程申请可用资源时，系统必须决策批准该申请后的系统是否还处于安全状态(Safe State)。进程序列(P_1, P_2, …, P_n)在当前分配状态下为安全序列(Safe Sequence)是指，对于每个P_i，它仍然可以申请的资源数小于当前可用资源加上所有进程P_j(其中$j<i$)所占用的资源总数。若P_i所需资源暂不可用，可以等待P_j进程结束释放该资源；当P_j结束时，P_i可以获得所等待资源，继续运行直至结束，最后返还所占用资源；当P_i结束时，P_{i+1}可以继续获得P_i释放的资源……直到所有进程都正常结束。如果存在一个安全序列，那么系统就处于安全状态，如果找不到任何一个安全序列，那么系统状态就是非安全的(Unsafe)。若系统处于安全状态，则不会出现死锁。相反，系统在非安全状态下可能会出现死锁(如图6-7所示)。死锁避免算法就是保证系统不会进入非安全状态。

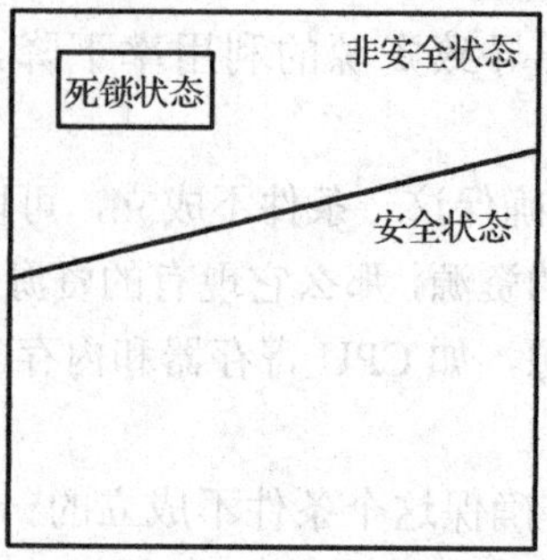

图 6-7　安全状态、非安全状态和死锁状态的状态空间

假设一个系统有 12 台磁带驱动器和 3 个进程 P_0、P_1 及 P_2。对磁带驱动器，进程 P_0 的最大需求为 10 台，P_1 的最大需求为 4 台，P_2 的最大需求为 9 台。假设，在 t_0 时刻，进程 P_0 占用 5 台磁带驱动器，进程 P_1 占用 2 台磁带驱动器，进程 P_2 占用 2 台磁带驱动器(还有 3 台空闲磁带驱动器)，情况如下。

进程	最大需求	当前占用	剩余需求
P_0	10	5	5
P_1	4	2	2
P_2	9	2	7

在 t_0 时刻系统处于安全状态，因为我们可以找到一个安全序列(P_1, P_0, P_2)。由于系统中还剩余 3 台空闲磁带驱动器，进程 P_1 的剩余需求量为 2，可以立即分配到所有的磁带驱动器，运行结束后返还它们，此时磁带驱动器剩余量为 5；接着，进程 P_0 可以得到所有的磁带驱动器，再返还它们，系统就会有 10 台可用的磁带驱动器；最后，进程 P_2 可以得到所有磁带驱动器，再返还它们，系统可用磁带驱动器数量会恢复到 12，所有进程运行结束。

但是系统也可能从安全状态转为非安全状态。假设进程 P_2 申请且得到 1 台磁带驱动器，那么系统资源的分配情况就变为如下情况(还有 2 台空闲磁带驱动器)。

进程	最大需求	当前占用	剩余需求
P_0	10	5	5
P_1	4	2	2
P_2	9	3	6

这时，只有进程 P_1 能得到所有磁带驱动器。当它返还资源时，系统只有 4 台磁带驱动器可用，已无法满足 P_0 和 P_2 进程的剩余需求，导致这两个进程处于无限等待状态，即发生死锁。这样我们就找不到任何安全序列，系统也因此处于不安全状态。为了避免这个情况，系统应当不批准之前 P_2 申请的 1 台磁带驱动器，严格按照(P_1, P_0, P_2)安全序列依次分配磁带驱动器。

6.5.2　银行家算法

银行家算法(Banker’s Algorithm)由 Dijkstra 提出，通过模拟“进程-资源”分配测试系统的安全状态，从而决定是否批准进程的资源申请。算法要求：新进程进入系统时应声明可能需要的每种资源类型的实例的最大数量，当进程申请一组资源时，算法模拟资源分配并确定这些资源的分配是否仍会使系统处于安全状态。如果会，就分配资源；否则，进程应等待，直到某个其他进程释放足够多的资源为止。

为了实现银行家算法，需要有几种数据结构，这些数据结构对资源分配系统的状态进行记录。这里，n 为系统进程的数量，m 为资源类型的种类。

Available：是一个长度为 m 的向量，表示每种资源类型的可用实例数量。

Max：是一个 $n \times m$ 的矩阵，定义每个进程的最大需求。

Allocation：是一个 $n \times m$ 的矩阵，定义每个进程现在被分配的每种资源类型的实例数量。

Need：是一个 $n \times m$ 的矩阵，表示每个进程还需要的剩余资源。

假设有一个系统，它有 5 个进程：T_0、T_1、T_2、T_3 及 T_4，3 种资源类型：A、B 及 C（n=5，m=3）。资源类型 A 有 10 个实例，资源类型 B 有 5 个实例，资源类型 C 有 7 个实例。图 6-8 展示了 t_0 时刻 Max 矩阵、Allocation 矩阵、Need 矩阵、Available 向量的内容。Need 是 Max 减去 Allocation 的结果，Available=(3,3,2)，也就是说 A、B、C 类资源分别剩余 3 个、3 个及 2 个。

Available		
A	B	C
3	3	2

	Max			Need			Allocation		
	A	B	C	A	B	C	A	B	C
T_0	7	5	3	7	4	3	0	1	0
T_1	3	2	2	1	2	2	2	0	0
T_2	9	0	2	6	0	0	3	0	2
T_3	2	2	2	0	1	1	2	1	1
T_4	4	3	3	4	3	1	0	0	2

图 6-8　t_0 时刻银行家算法矩阵图

我们尝试找出一个进程的安全序列，根据 Need 矩阵和 Available 向量可知，现有的系统剩余资源只能满足 T_1 和 T_3 的要求，我们先假设满足 T_1，那么 T_1 就可以运行结束然后将 Allocation 的 T_1 行向量 (2,0,0) 返还系统，Available=(3,3,2)+(2,0,0) = (5,3,2)；再用类似的方法寻找下一个可满足要求的进程，完整的过程如图 6-9 所示。这样我们就找到了一个安全序列{T_1,T_3,T_4,T_2,T_0}，因此系统在此时刻处于安全状态。

Step	Available A	B	C	Done	Choice
1	3	3	2	—	T_1
2	5	3	2	T_1	T_3
3	7	4	3	T_3	T_4
4	7	4	5	T_4	T_2
5	10	4	7	T_2	T_0
6	10	5	7	T_0	☆

图 6-9　安全序列

如果此时刻进程 T_1 发出新的请求 Request=(1,0,2)，系统会再次使用银行家算法检测安全状态。算法会先假设满足要求，则 Available = (3,3,2) − (1,0,2) = (2,3,0)，再用相同的方法尝试找出一个进程安全序列，最后的安全序列为{ T_1, T_0, T_4, T_3, T_2}，系统也就可以批准这个请求。

用同样的方法可以计算出，在 t_0 时刻不能批准 T_4 的请求 Request=(3,3,0)，但是可以批准 T_0 的请求 Request=(0,2,0)，计算过程留给读者自行完成。

6.6 死锁的检测和恢复

死锁避免算法最大的问题在于系统必须事先知道进程运行过程中所需资源的最大数量，但是在实践中，这些信息的获取是有难度的或者是精度不高的，这给银行家算法的执行带来巨大的挑战。即使银行家算法可以执行，为了保证系统始终处于安全状态，也会要求并发进程严格按照安全序列的顺序运行，这又给进程的同步带来了问题。因此，本节将介绍一种死锁的检测方法，它允许死锁的出现，但是死锁可以被算法检测出来，并讨论一些可以从死锁状态恢复出来的算法。

6.6.1 死锁的检测

死锁检测（Deadlock Detection）的方法是对资源分配图进行简化，如果可以将资源分配图完全简化，则说明系统中没有死锁，否则有死锁。

考虑图 6-10 所示的例子，简化过程如下。

(1) 在资源分配图中，找出一个既不阻塞又非独立的进程节点 P_i。在顺利的情况下，P_i 可获得所需资源而继续运行，直至运行完毕，再释放其所占用的全部资源，这相当于消去 P_i 的请求边和分配边，使之成为孤立节点。在图 6-10 中，将 P_2 和 P_4 的两个分配边消去后，它们成为孤立节点。

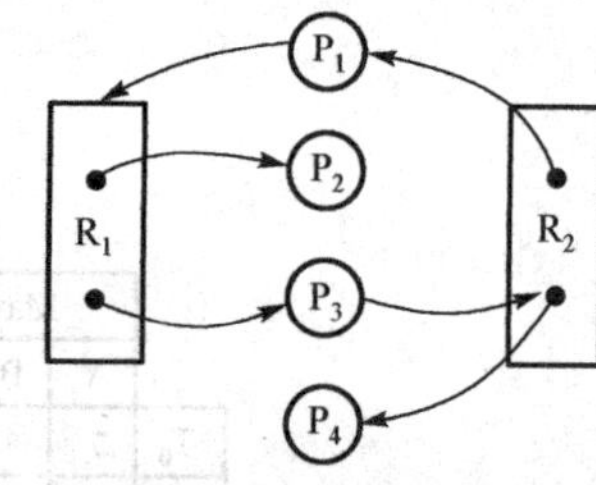

图 6-10 资源分配图的简化

(2) P_2 释放资源后，便可使 P_1 获得资源而继续运行，直至 P_1 运行完成后又释放出它所占用的 R_2，P_1 的分配边及申请边也可以消去，成为孤立节点。

(3) 用同样的方法可以将 P_3 也变为孤立节点。

(4) 在进行一系列的简化后，若能消去图中所有的边，使所有的节点都成为孤立节点，则称该图是可完全简化的；若不能通过任何过程使该图完全简化，则称该图是不可完全简化的。

系统状态为死锁状态的充分条件是，系统状态的资源分配图是不可完全简化的。该定理称为死锁定理。

6.6.2 死锁的恢复

当检测算法确定已有死锁时，存在多种可选方案来消除死锁。一种方案是，通知程序员死锁已发生，并且需人工处理死锁。另一种方案是，让系统从死锁状态中自动恢复（Recover）。

1. 进程终止

通过终止处于死锁状态的进程可以强行回收它们占用的资源从而解除死锁。可以终止所有死锁进程，也可以终止部分死锁进程，但两种方法的代价都是比较大的。系统有时要有选择性地挑选代价最小进程让其终止。

2. 资源抢占

通过资源抢占来消除死锁，就是不断地抢占一些进程的资源以便给其他进程使用，直到死锁解除为止。如果要采用资源抢占来处理死锁，那么需要处理以下三个问题。

(1)选择牺牲进程：抢占哪些资源和哪些进程？与进程终止一样，应确定抢占的顺序，使代价最小。代价因素包括死锁进程拥有的资源数量、死锁进程到现在为止所消耗的时间等。

(2)回滚：如果从一个进程那里抢占了一个资源，那么应对该进程做什么安排？显然，该进程不能继续正常运行；它缺少所需的某些资源，应将该进程回滚到某个安全状态，以便从该状态重启进程。因为一般来说，很难确定什么是安全状态，所以最简单的解决方案是完全回滚：终止进程并重新运行。然而，更有效的方法是回滚进程直到足够打破死锁，但是这种方法要求系统维护有关进程运行的更多状态信息。

(3)饥饿：如何确保不会发生饥饿现象？即如何保证资源不会总从同一个进程中被抢占。如果一个系统是基于代价来选择牺牲进程的，那么同一个进程可能总被选为牺牲进程，这个进程永远不能完成指定任务，因为任何实际系统都需要处理这种饥饿情况。显然，应确保一个进程只能有限次数地被选为牺牲进程，最常用的方法是在代价因素中加上回滚次数。

6.7 Reading Materials

6.7.1 Overview

A deadlocked state occurs when two or more processes are waiting indefinitely for an event that can be caused only by one of the waiting processes. There are three principal methods for dealing with deadlocks:

- Use some protocol to prevent or avoid deadlocks, ensuring that the system will never enter a deadlocked state.
- Allow the system to enter a deadlocked state, detect it, and then recover.
- Ignore the problem altogether and pretend that deadlocks never occur in the system.

The third solution is the one used by most operating systems, including Linux and Windows.

A deadlock can occur only if four necessary conditions hold simultaneously in the system: mutual exclusion, hold and wait, no preemption, and circular wait. To prevent deadlocks, we can ensure that at least one of the necessary conditions never holds.

A method for avoiding deadlocks, rather than preventing them, requires that the operating system have a priori information about how each process will utilize system resources. The banker's algorithm, for example, requires a priori information about the maximum number of each resource class that each process may request. Using this information, we can define a deadlock-avoidance algorithm.

If a system does not employ a protocol to ensure that deadlocks will never occur, then a detection-and-recovery scheme may be employed. A deadlock-detection algorithm must be invoked to determine whether a deadlock has occurred. If a deadlock is detected, the system must recover either by terminating some of the deadlocked processes or by preempting resources from some of the deadlocked processes.

Where preemption is used to deal with deadlocks, three issues must be addressed: selecting a victim, rollback, and starvation. In a system that selects victims for rollback primarily on the basis of cost factors, starvation may occur, and the selected process can never complete its designated task.

Researchers have argued that none of the basic approaches alone is appropriate for the entire spectrum of resource-allocation problems in operating systems. The basic approaches can be combined, however, allowing us to select an optimal approach for each class of resources in a system.

6.7.2 Dijkstra Biography

Born in Rotterdam, Netherlands, Edsger Dijkstra studied theoretical physics at Leiden University, but he quickly realized he was more interested in computer science. Originally employed by the Mathematisch Centrum in Amsterdam, he held a professorship at the Eindhoven University of Technology in the Netherlands, worked as a research fellow for Burroughs Corporation in the early 1970s, and later held the Schlumberger Centennial Chair in Computer Sciences at The University of Texas at Austin, in the United States. He retired in 2000.

Among his contributions to computer science is the shortest path-algorithm, also known as Dijkstra's algorithm; Reverse Polish Notation and related Shunting yard algorithm; the THE multiprogramming system; Banker's algorithm; and the semaphore construct for coordinating multiple processors and programs. Another concept due to Dijkstra in the field of distributed computing is that of self-stabilization - an alternative way to ensure the reliability of the system. Dijkstra's algorithm is used in SPF, Shortest Path First, which is used in the routing protocol OSPF, Open Shortest Path First.

While he had programmed extensively in machine code in the 1950s, he was known for his low opinion of the GOTO statement in computer programming, writing a paper in 1965, and culminating in the 1968 article "A Case against the GO TO Statement" (EWD215), regarded as a major step towards the widespread deprecation of the GOTO statement and its effective replacement by structured control constructs, such as the while loop. This methodology was also called structured programming, the title of his 1972 book, coauthored with C.A.R. Hoare and Ole-Johan Dahl. The March 1968 ACM letter's famous title, "Go To Statement Considered Harmful," was not the work of Dijkstra, but of Niklaus Wirth, then editor of Communications of the ACM.

Dijkstra was known to be a fan of ALGOL 60, and worked on the team that implemented the first compiler for that language. Dijkstra and Jaap Zonneveld, who collaborated on the compiler, agreed not to shave until the project was completed.

Dijkstra wrote two important papers in 1968, devoted to the structure of a multiprogramming operating system called THE, and to Co-operating Sequential Processes.

He is famed for coining the popular programming phrase "2 or more, use a for", alluding to the fact that when you find yourself processing more than one instance of a data structure, it is time to encapsulate that logic inside a loop.

From the 1970s, Dijkstra's chief interest was formal verification. The prevailing opinion at the time was that one should first write a program and then provide a mathematical proof of correctness. Dijkstra objected noting that the resulting proofs are long and cumbersome, and that the proof gives no insight as to how the program was developed. An alternative method is program derivation, to "develop proof and program hand in hand." One starts with a mathematical specification of what a program is supposed to do and applies mathematical transformations to the specification until it is turned into a program that can be executed. The resulting program is then known to be correct by construction. Much of Dijkstra's later work concerns ways to streamline mathematical argument. In a 2001 interview he stated a desire for "elegance", whereby the correct approach would be to process thoughts mentally, rather than attempt to render them until they are complete. The analogy he made was to contrast the compositional approaches of Mozart and Beethoven.

Dijkstra was one of the very early pioneers of the research on distributed computing. Some people even consider some of his papers to be those that established the field. In particular, his paper "Self-stabilizing Systems in Spite of Distributed Control" started the sub-field of self-stabilization.

Dijkstra was known for his essays on programming; he was the first to make the claim that programming is so inherently difficult and complex that programmers need to harness every trick and abstraction possible in hopes of managing the complexity of it successfully.

Dijkstra believed that computer science was more abstract than programming; he once said, "Computer Science is no more about computers than astronomy is about telescopes."

He died in Nuenen, The Netherlands on August 6, 2002 after a long struggle with cancer. The following year, the ACM（Association for Computing Machinery）PODC Influential Paper Award in distributed computing was renamed the Dijkstra Prize in his honour.

第7章 内存管理

7.1 概述

7.1.1 基本概念

现代计算机采用层次结构的存储系统，以便在容量大小、速度快慢及价格高低等因素中取得平衡点，获得较好的性能价格比。如图 7-1 所示，计算机的存储层次结构呈一个金字塔形状，越往上存储介质的访问速度越快，但存储容量越小，其价格也会越高。顶层的寄存器(Register)是速度最快但最昂贵的存储器，容量非常小，通常以字(Word)为单位。高速缓存(Cache)是一种存取速度比主存快，但容量比主存小得多的存储器，利用它存储主存中一些经常访问的数据，可以大幅提高程序运行速度，加快访问主存的相对速度。高速缓存也分级别，级别越低，离 CPU 就越近，速度越快、容量越小。

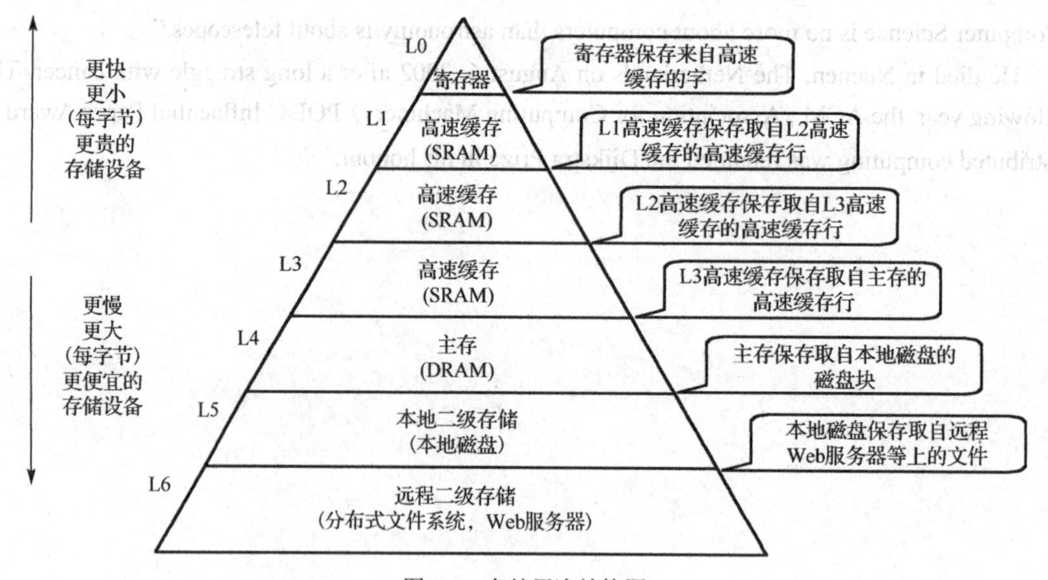

图 7-1　存储层次结构图

主存(Main Memory)是现代计算机系统运行的中心，如图 7-2 所示。很多时候人们习惯称为“内存”(Memory)，即“内部存储器”，其和“外存”，即“外部存储器”(Storage)正好对应。内存是由很大的一组字节构成的，每字节都有唯一的硬件地址。处理器和 I/O 设备之间及 I/O 设备之间传输数据必须经由内存中转，例如，向显示器输出字符串“hello world”，该字符串数据先存在内存中，再经总线(Bus)传输至显示器。再如，键盘上输入的字符都暂时保存在内存中。所有的程序必须进入内存成为进程后方可运行，进程在内存中的实体依靠 PCB 进行管理，CPU 根据 PC 寄存器的值从内存中取出下一条要执行的指令，指令接着被译码执行。同时可能需要从内存中读取操作数，指令执行的结果可能会被存回内存。

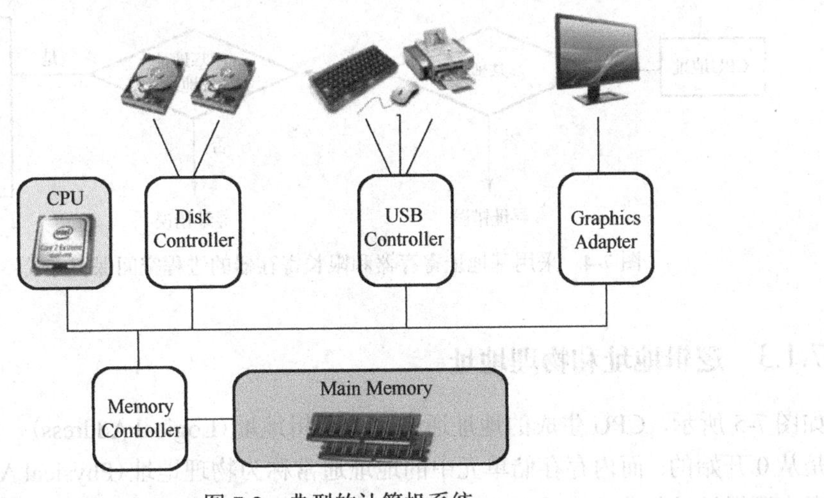

图 7-2　典型的计算机系统

支持多道程序设计的计算机系统允许进程并发运行，在内存中同时装有多道进程的实体，同时操作系统代码也在内存中。如何隔离用户进程、操作系统、用户进程及用户进程，保证用户进程不可以访问操作系统的内存空间，同时保证用户进程之间不会进行未授权访问，这是系统设计人员需要考虑的重要问题。

7.1.2　基本硬件

我们不仅要关心访问物理内存的相对速度，而且要确保操作正确。我们应保护操作系统不被用户进程访问。在多用户操作系统上，我们还应保护用户进程不会互相影响。由于操作系统通常不干预 CPU 对内存的访问，因此这种保护通过硬件来实现。

首先，我们需要确保每个进程都有一个独立的内存空间。独立的进程空间可以保证进程之间不互相影响，这对于将多个进程加入内存以便并发运行来说至关重要。如图 7-3 所示，有三个独立的进程空间和操作系统空间。左侧的数字表示的是内存的硬件地址，从 0 开始编址，每字节都有个唯一的地址。操作系统装载的位置通常是固定的，有些操作系统会从低地址开始装载，有些则从高地址开始装载，本书不做严格的规定，两种方法都会采用。

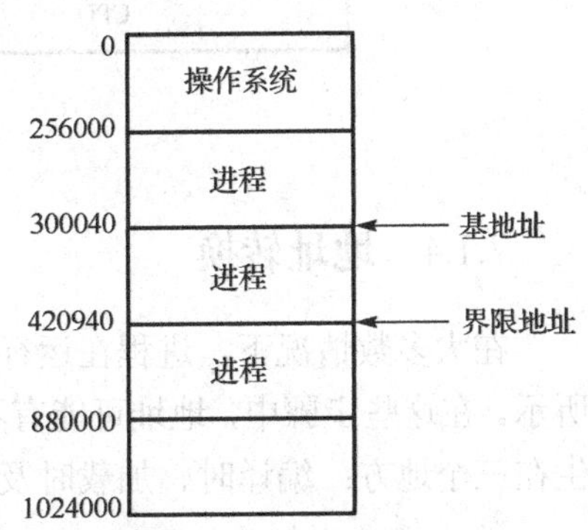

图 7-3　独立的进程空间

为了实现内存空间的保护，需要两个额外的硬件寄存器：基地址寄存器(Base Register, BR)和限长寄存器(Limit Register, LR)。基地址寄存器保存进程的起始地址，而限长寄存器保存进程的长度，这两个寄存器的值只能被操作系统的特权指令加载，这种方式允许操作系统修改这两个寄存器的值，但不允许用户进程修改它们。比如，图 7-3 中的基地址为 300040，限长为 120900，表示合法访问的地址范围为从 300040 到 420939(含)。

使用这两个寄存器实施进程空间保护的过程如图 7-4 所示。小于基地址或是大于基地址和限长地址之和的内存地址访问都会被禁止。

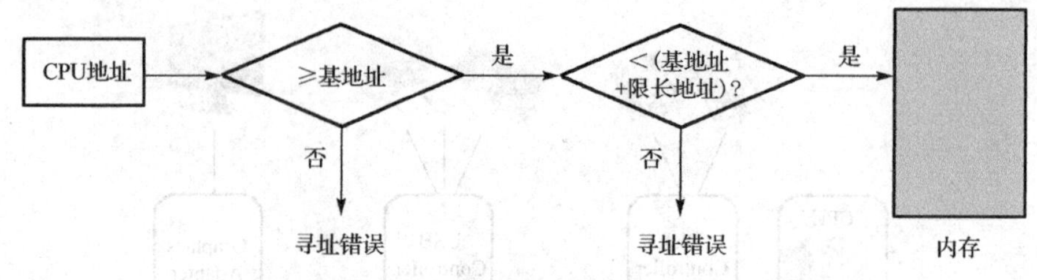

图 7-4　采用基地址寄存器和限长寄存器的进程空间保护方式

7.1.3　逻辑地址和物理地址

如图 7-5 所示，CPU 生成的地址通常称为逻辑地址(Logical Address)，每个进程的逻辑地址都是从 0 开始的。而内存存储单元中的地址通常称为物理地址(Physical Address)。由程序生成的所有逻辑地址的集合称为逻辑地址空间(Logical Address Space)，这些逻辑地址对应的所有物理地址的集合称为物理地址空间(Physical Address Space)。

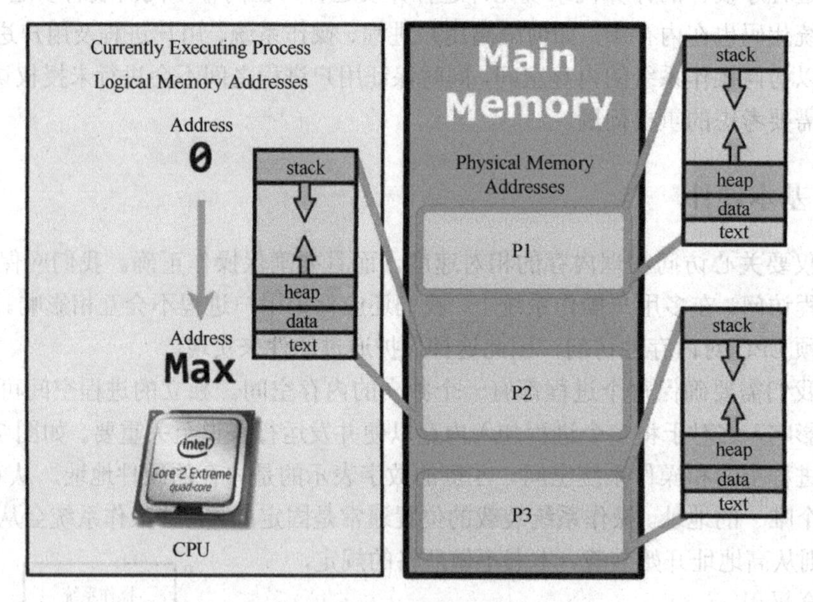

图 7-5　逻辑地址和物理地址

7.1.4　地址转换

在大多数情况下，进程在运行前，需要经过多个步骤，其中有的步骤是可选的，如图 7-6 所示。在这些步骤中，地址可能有不同的表示形式。程序的逻辑地址和物理地址的绑定可以发生在三个地方：编译时、加载时及运行时。

1. 编译时(Compile Time)

如图 7-7 所示，图 7-7(a)左侧是生成好逻辑地址的程序，右侧是内存。如果在编译时就能知道进程在内存中的驻留地址，那么就可以直接生成绝对代码(Absolute Code)，即物理地址。例如，如果事先就知道用户进程驻留在内存地址 R 处，那么生成的编译代码地址就可以从该位置开始向后延伸[如图 7-7(b)所示]。这种绑定方式的好处是，加载内存后访问的逻辑地址即物理地址，省去了转换步骤，访问效率加快，但缺点是不够灵活，因为程序在内存中的物理位置已经固定，如果起始地址发生变化，那么就必须重新编译代码。

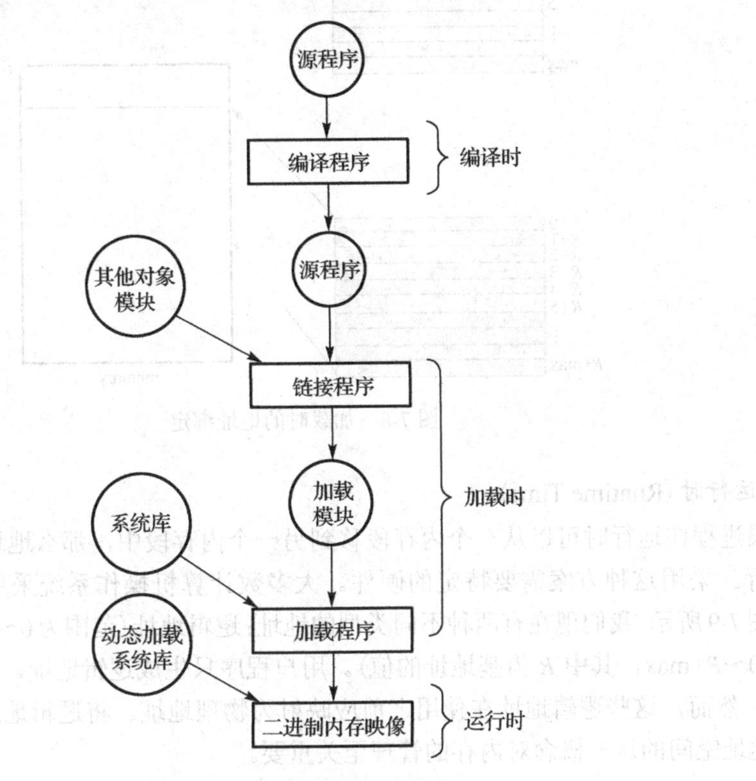

图 7-6　进程运行前的多步骤处理

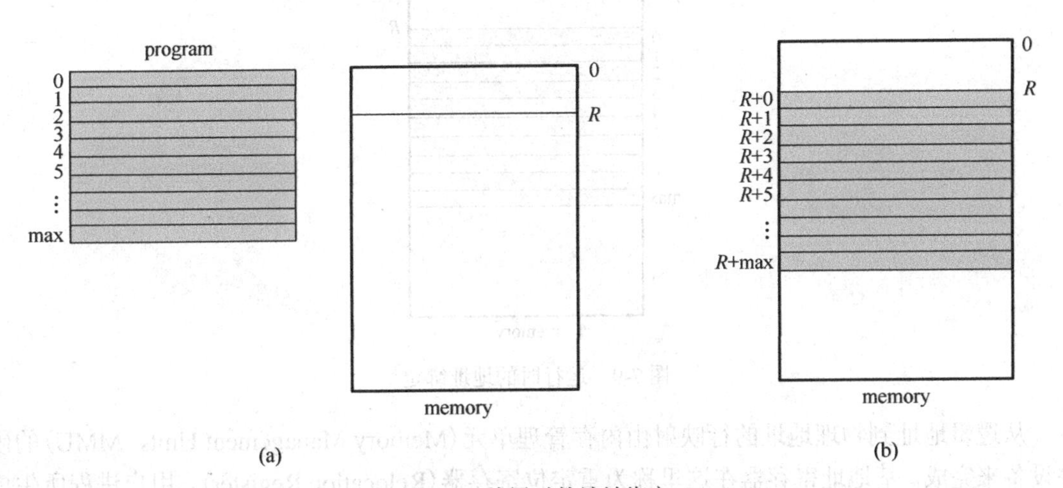

图 7-7　编译时的地址绑定

2. 加载时(Load Time)

如图 7-8 所示，如果在编译时并不知道进程将驻留在何处，那么编译器就应生成可重定位代码(Relocatable Code)，地址绑定会延迟到加载时才进行。如果起始地址发生变化，那么不需要重新编译代码，只需重新生成可重定位代码即可。

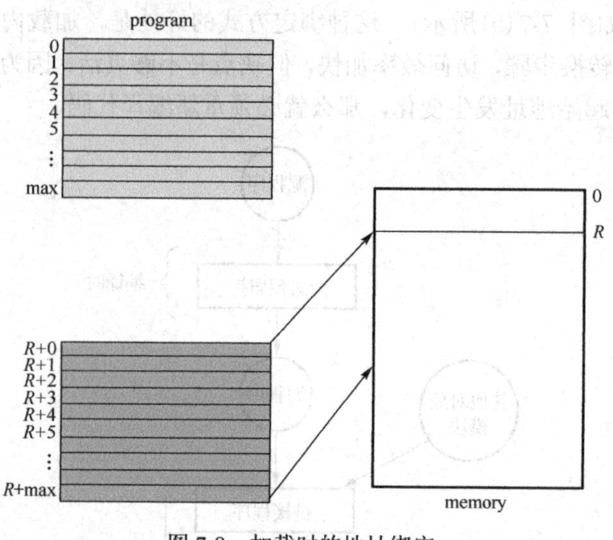

图 7-8　加载时的地址绑定

3. 运行时(Runtime Time)

如果进程在运行时可以从一个内存段移到另一个内存段中，那么地址绑定应延迟到运行时才进行。采用这种方案需要特定的硬件。大多数计算机操作系统采用这种方案。

如图 7-9 所示，我们现在有两种不同类型的地址：逻辑地址（范围为 0～max）和物理地址（范围为 R+0～R+max，其中 R 为基地址的值）。用户程序只生成逻辑地址，且程序的地址空间为 0～max。然而，这些逻辑地址在使用之前应映射为物理地址。将逻辑地址空间绑定到另一单独物理地址空间的这一概念对内存的管理至关重要。

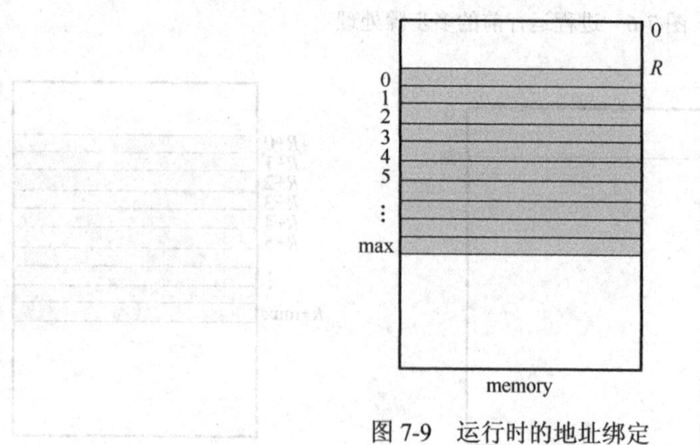

图 7-9　运行时的地址绑定

从逻辑地址到物理地址的行映射由内存管理单元(Memory Management Unit，MMU)的硬件设备来完成。基地址寄存器在这里称为重定位寄存器(Relocation Register)，用户进程所生成的地址在被送入内存之前，都将加上重定位寄存器的值，如图 7-10 所示。例如，假设基地址为 14000，那么用户对逻辑地址 0 的访问将动态地重定位为对物理地址 14000 的访问；对逻辑地址 346 的访问将动态地重定位为对物理地址 14346 的访问。

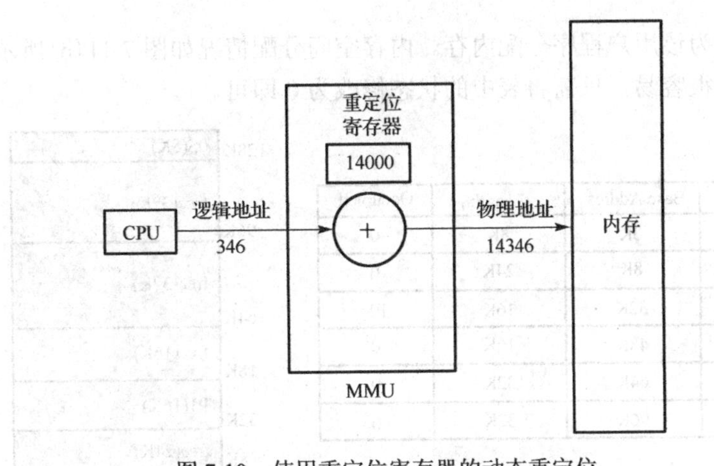

图 7-10　使用重定位寄存器的动态重定位

用户进程看不到真实的物理地址，进程处理的地址都是逻辑地址，只有当逻辑地址作为内存地址时，它才会和重定位寄存器的值相加进行重定位转换得到物理地址。

7.2　连续内存分配

现在我们要讨论如何使用内存空间，内存作为一个仓库，应当尽可能地被高效利用，本节介绍一种常用方法——连续内存分配(Contiguous Memory Allocation)。这种方法要求每个进程在内存中的区域必须是连续的，本节将介绍两种连续分配方式：固定分区分配(Fixed-sized Partion)和可变分区分配(Variable Partion)，无论介绍哪种方式，我们都将讨论三个问题：分配方法、内存回收方法及内存保护方法。

7.2.1　固定分区分配

固定分区式分配事先将内存空间划分为若干固定大小的分区，每个分区中只装入一个作业，新作业到达时，系统会在所有空闲分区中找一个可容纳该作业的最小分区分配给它，系统会使用一张表格记录分区的空闲和分配状况。

1. 分配方法

(1) 分区大小相等。其缺点是缺乏灵活性，即当作业太小时，会造成内存空间的浪费；当作业太大时，一个分区又不足以装入该程序，致使作业无法运行。这种分配方法适用于运行作业所需内存空间大小相等的情形，例如，炉温群控系统就利用一台计算机控制多台相同的冶炼炉。

(2) 分区大小不等。为了克服分区大小相等缺乏灵活性的缺点，可把内存区划分成多个较小分区、适量的中等分区及少量大分区。这样，便可根据作业的大小为其分配适当的分区。

2. 内存回收方法

为了便于内存分配，通常将分区按大小进行排列，并为其建立一张分区使用表，表项包括每个分区的分区号、起始地址、大小及状态(Occupied，0 表示空闲)，如图 7-11(a)所示。当有作业要装入时，由内存分配程序检索该表，从中找出一个能满足要求且尚未分配的分区，将其分配给该程序，然后将该表项的状态设置为作业编号，表示“已分配”；若未找到大小足够的分区，则拒绝为该用户程序分配内存。内存空间分配情况如图 7-11(b)所示。当作业运行结束，回收空间也很容易，只需将表中的状态修改为 0 即可。

No.	Base Address	Length	Occupied
1	0K	8K	0
2	8K	24K	0
3	32K	16K	P1
4	48K	16K	0
5	64K	32K	0
6	96K	32K	0

(a)

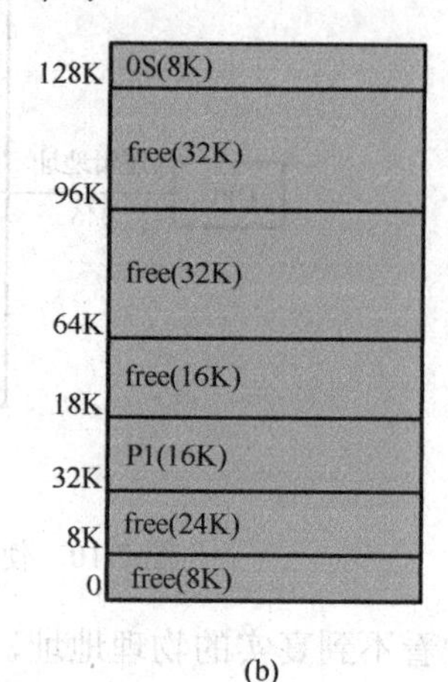

(b)

图 7-11　固定分区分配

3. *内存保护方法*

固定分区分配的内存保护可采用 7.1.2 节中所述的方案，配合基地址寄存器及限长寄存器进行。在图 7-11(b)所示的分配方案中，P1 的基地址是 32K，限长是 16K，如果请求的逻辑地址超出了[32K, 48K–1]的范围，就会发出地址请求错误中断。

7.2.2　可变分区分配

采用可变分区分配方案时，对内存不进行事先的划分，而随着新进程到达动态地寻找可用空间进行分配。为了实现这个功能，操作系统要分别记录未分配的内存起始地址和大小，以及已分配的内存起始地址和大小。

在初始阶段，所有内存都是空闲的，这种一大块的可用内存空间称为孔(Hole)。随着进程内存的分配和释放，孔的可用数量和大小也会随着发生变化。如图 7-12 所示，图 7-12(a)说明内存的所有空间都被占满，当 Process 8 终止，空间被回收后，就会出现一个孔，如图 7-12(b)所示。当将其分配给新到达的进程 Process 9 后，孔变小了，如图 7-12(c)所示。类似地，可以继续分配给 Process 10。通常，可用的内存块分散在内存中的不同大小的孔集合中。当新进程需要内存时，操作系统为该进程查找足够大的孔。如果孔太大，那么就分为两块：一块分配给新进程，另一块依旧回到孔集合中。当进程终止时，内存将被释放，该内存将依旧回到孔集合中。如果新孔与其他孔相邻，那么将这些孔合并成大孔。这时，系统可以检查是否有进程在等待内存空间，以及新合并的内存空间是否满足等待进程等。

当内存中孔的数量不止一个(如图 7-13 所示)，且都可以满足新进程的大小需求时，操作系统有三种孔的选择方案：首次适应(First-Fit)、最优适应(Best-Fit)及最差适应(Worst-Fit)。

(1) 首次适应：分配首个足够大的孔。查找可以从头开始，也可以从上次首次适应结束时开始。一旦找到足够大的空闲孔，就可以停止。

(2) 最优适应：分配最小的足够大的孔。应查找整个列表，除非列表按大小排序。这种方法可以产生最小剩余孔。

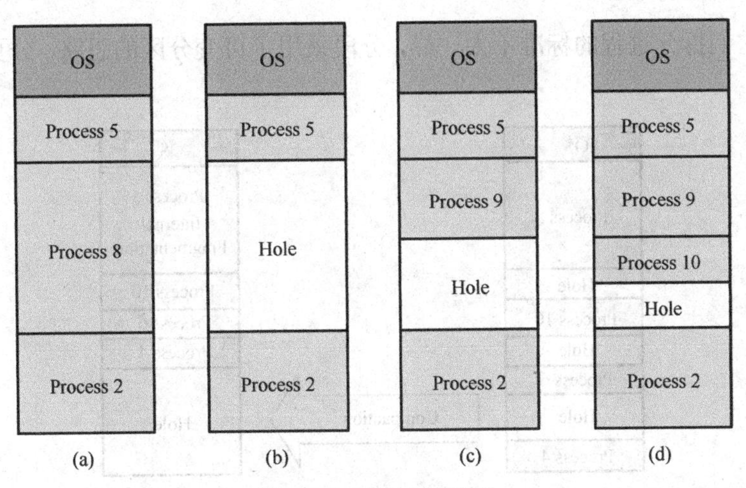

图 7-12 可变分区分配

(3) 最差适应：分配最大的孔。同样应查找整个列表，除非列表按大小排序。这种方法可以产生最大剩余孔，该孔可能比最优适应产生的最小剩余孔更实用。

首次适应和最优适应在执行时间和空间利用方面都优于最差适应。首次适应和最优适应在空间利用方面难分伯仲，但是首次适应要更快些。

可变分区分配的内存保护方法和固定分区分配是类似的，因为进程在内存中都是连续存放的，所以只要知道进程加载的起始地址和进程长度就可以实现内存保护。

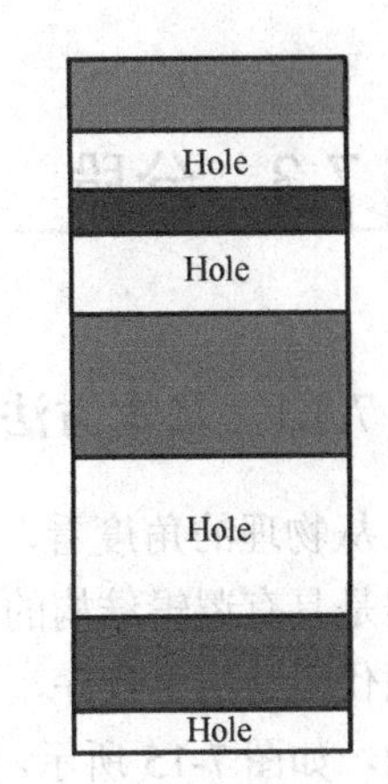

图 7-13 动态存储分配

7.2.3 碎片

碎片 (Fragmentation) 是内存中一些非常小且难以被利用的空闲区域。随着内存空间的分配和回收，碎片的产生是不可避免的，不同的分区分配方法产生的碎片也是不一样的。

对于固定分区分配方案而言，每个作业占用一个分区，分区的大小是事先划分好的，作业的大小不太可能正好和分区一样，通常都小于分区大小，这样该分区内未使用的部分也不可以被其他作业使用，从而成为"内部碎片"(Internal Fragmentation)。在可变分区分配方案中，随着进程被加载到内存中和从内存中退出，空闲内存空间被分为大量的孔，就出现了"外部碎片"(External Fragmentation) 问题。最坏情况下，每两个进程之间就会有一块碎片。

解决外部碎片问题的一种方法是紧缩 (Compaction)，如图 7-14 所示。它通过移动内存数据，让所有的碎片得以连续，从而合并成一个大的新孔。然而，紧缩不总是可以做的：

(1) 对于静态地址重定位，紧缩是不可行的，因为它的地址在运行前就已经固定，不可以在内存中移动；

(2) 紧缩方法要在内存中移动进程，在实施前要评估移动的开销。

如果换个思路，我们会发现碎片是很小且无法被利用的空间。假设进程不要求连续存放，是不是可以把进程拆散？这样就不需要将碎片合并，而是将拆散的进程离散地插到这些零星的碎片里。从 7.3 节开始我们介绍两种方案：分段和分页。两种方案的核心思路是将进程进行离散存放，但是它们拆分进程的标准不太一样，分段采用了可变分区的思路，分页则采用了固定分区的思路。

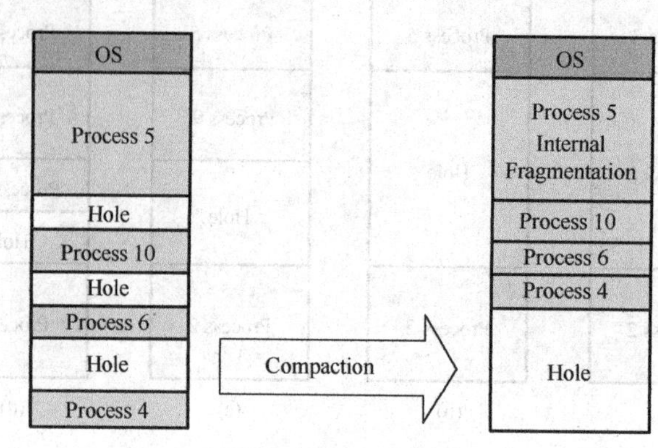

图 7-14　紧缩

7.3　分段

获取视频

7.3.1　基本方法

从物理的角度看，内存仅仅是一组字节的集合，而在用户（开发人员）的眼中，内存中的数据是具有逻辑结构的。这个结构将数据分成了若干大小不等的区域（段落），就如同一段程序的源代码是由主程序、子程序及函数等部分构成的，还有对象、数组、堆栈、变量及符号表等结构，如图 7-15 所示。

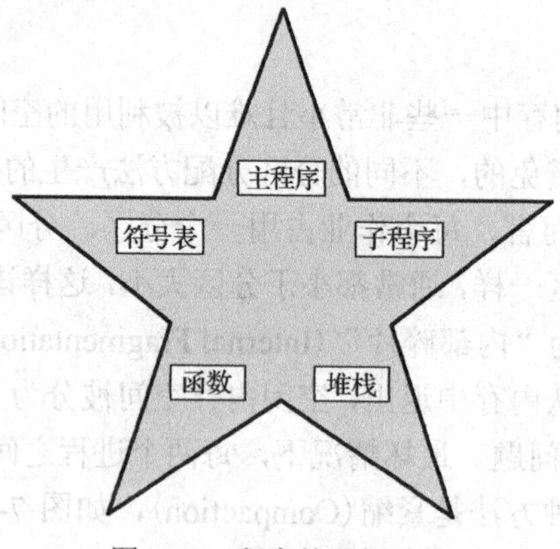

图 7-15　程序的逻辑结构

分段（Segmentation）就是以用户视角将进程划分成若干逻辑段的内存管理方案，每段都有专门的用途。我们用一个例子引入这种方案。如图 7-16 所示，假设有这样的程序，左侧的序号对应着该程序的逻辑地址（从 0 开始），为了便于讨论，不给出具体的代码，我们先以函数为单位来划分段，如地址 0～4 是 f() 函数段、地址 5～8 是 g() 函数段、地址 9～13 是 main() 函数段。按照不连续分配的原则，将该程序切分成 3 段然后分别存入内存起始地址为 X、Y 及 Z 的地方，如图 7-17 所示。

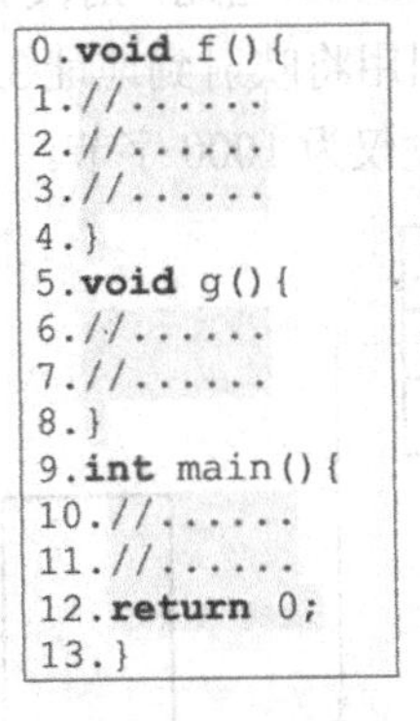

图 7-16 段的划分

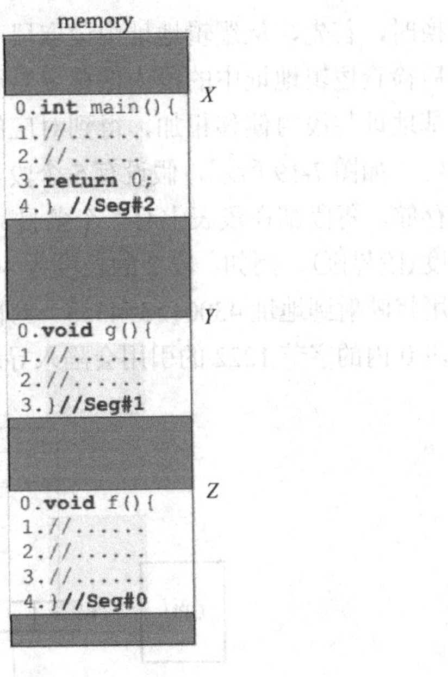

图 7-17 分段存储

注意，段在内存中的顺序和逻辑顺序无关，将原程序划分成为 3 段后，每段内的逻辑地址均是从 0 开始的，图 7-17 中，“Seg#”后面的数字代表段的编号，也是从 0 开始的。我们可归纳出以下信息：

(1) 程序被分成 3 段，段号分别是 0、1、2；

(2) 第 0 段加载的起始地址为 *Z*，第 1 段加载的起始地址为 *Y*，第 2 段加载的起始地址为 *X*；

(3) 第 0 段的段内逻辑地址为 0～4，第 1 段的段内逻辑地址为 0～3，第 2 段的段内逻辑地址为 0～4。

现在尝试计算一下图 7-16 中逻辑地址 7 所对应的物理地址是多少。逻辑地址 7 所在的段是 g() 函数段，即第 1 段，分段后原逻辑地址在段内的逻辑地址变成了 2（段内偏移），同时它加载的物理内存起始地址是 *Y*，因此其对应的物理地址是：*Y*+2 。

为了实现分段内存管理，即逻辑地址空间是由一组段构成的，每个段都有编号和长度，地址指定了段号和段内偏移。因此用户通过两个量来指定地址：<段号, 段内偏移>。

通常，在编译用户程序时，编译程序会根据输入程序来自动构造段。一个 C 编译器可能会创建如下段代码：全局变量、堆（内存从堆上分配）、每个线程使用的栈及标准的 C 库。在编译时，链接的库可能被分配不同的段，加载程序时会装入这些段，并为它们分配段号。

7.3.2 实现原理

因为分段实施了离散存储，所以各个分段存放的位置必须被记录下来，因此需要一些数据结构和硬件帮助我们完成上述工作。其中段表（Segment Table）是一个数据结构，它记录了每段的段基地址（Segment Base）和段界限（Segment Limit）。段基地址包含该段在内存中的起始物理地址，而段界限指定该段的长度。

段表的使用如图 7-18 所示，逻辑地址由两部分组成：段号 *s* 和段内偏移 *d*。在进行地址转换时，首先，从逻辑地址中读取段号 s，用 s 在段表中索引到对应段的段基地址和段界限；然后检查逻辑地址中的段内偏移 d 是否超出了段界限，如果超出了，则报寻址错误；最后，将段基地址与段内偏移相加，得到对应的物理地址。

如图 7-19 所示，假设有 5 个段，按 0～4 来编号。各段按图 7-19(b)所示的物理内存空间来存储。每段都在段表中有一个条目，它包括段在物理内存中的起始地址(段基地址)和该段的长度(段界限)。例如，段 2 的长度为 400 字节，起始地址为 4300。因此，对段 2 内的字节 53 的引用将映射到地址 4300+53=4353。对段 3 内的字节 852 的引用将映射到地址 3200+852=4052。对段 0 内的字节 1222 的引用会陷入寻址错误，因为该段长度仅为 1000 字节。

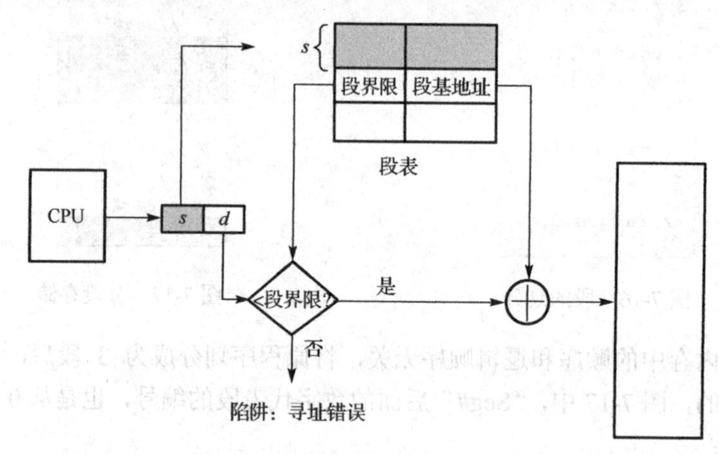

图 7-18　段表的使用

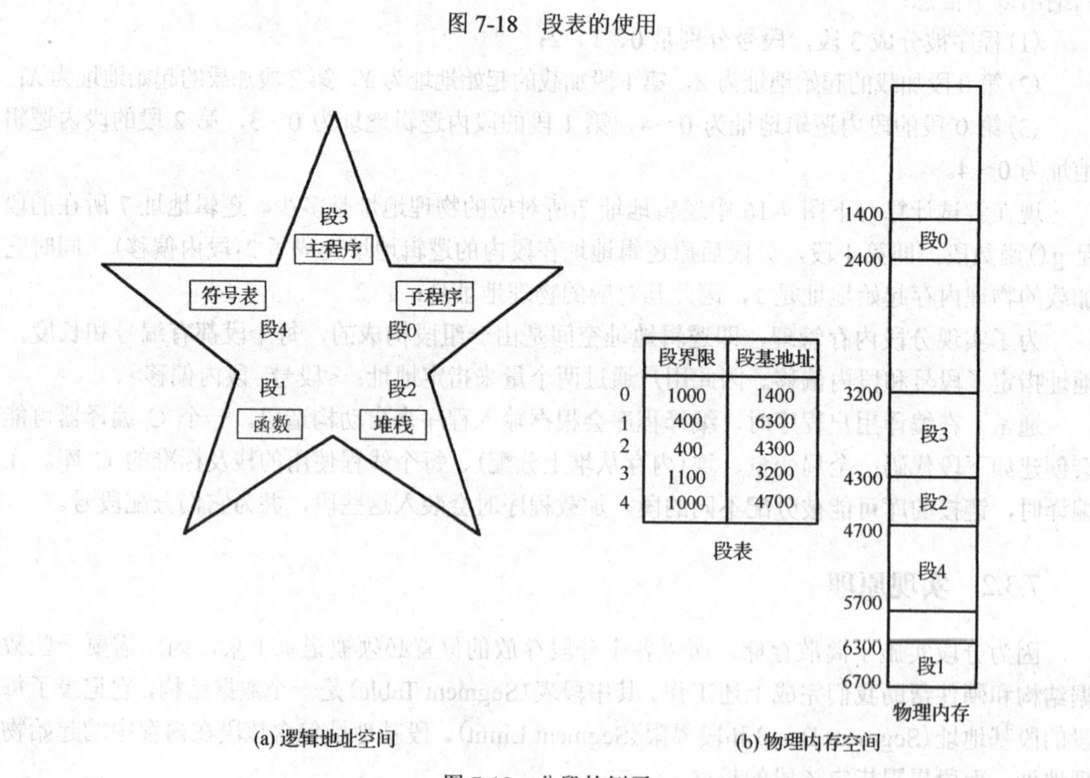

图 7-19　分段的例子

7.4 分页

7.4.1 基本方法

将物理内存空间分为固定大小的块，这些块称为页帧或页框(Frame)；将进程的逻辑内存也分为同样大小的块，这些块称为页或页面(Page)，如图 7-20 所示。页框和页面的大小是 2 的 *n* 次方，那么内存一定可以划分出整数个页框大小，但是进程的逻辑地址空间不一定是页面大小的整数倍，进程的最后一个页面内容可能会少于页面大小。就如同分段方案那样，将进程的页面离散地存入内存的页框中。不妨想象一下，因为页面和页框是一样大的，所以除了最后一个页面，其他页面正好填满整个页框。和固定分区连续分配方案相比，分页方案把每个进程产生的内部碎片控制在一个页框以内。和分段方案类似，必须要记录离散存放的页面的位置，记录位置的这个专门的数据结构称为“页表”(Page Table)，其作用是记录页面加载的页框编号。图 7-21 是一个页表示意图，页表第 1 列页面编号(简称页号)在实现中可以省略，它默认和页表条目的下标是吻合的。

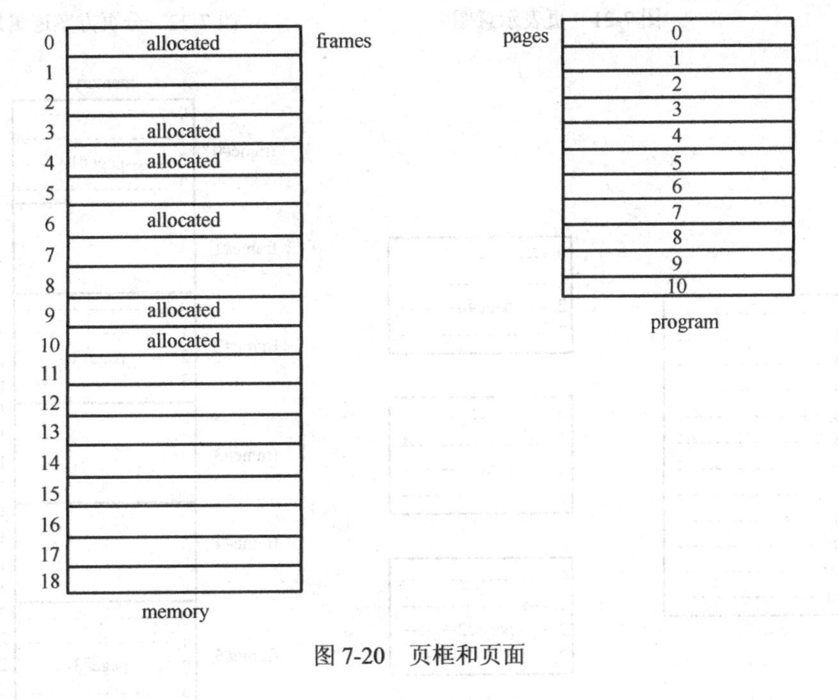

图 7-20 页框和页面

7.4.2 地址转换

分页方案中进程的逻辑地址被分成两部分：页号和页内偏移(如图 7-22 所示)。

我们从一个简单的例子开始讲起，图 7-23(a)指示的是一个进程的逻辑地址空间，为了方便讨论，假设每条指令只有 1 字节，逻辑地址空间从 0 开始，共 12 条指令。再假设页面大小为 2^2=4 字节，那么该进程可以被分成 3 个页面，每个页面的页内偏移均是 0～3，共 4 条指令，页号从上至下分别0、1及2[如图7-23(b)所示]。这3个页面分别存入了内存的第5、0及2号页框，如图7-24所示，图中最右侧的是内存的物理地址。

	memory
0	allocated
1	1
2	2
3	allocated
4	allocated
5	0
6	allocated
7	6
8	4
9	allocated
10	allocated
11	10
12	7
13	
14	3
15	9
16	5
17	
18	8

Page No.	Frame No.
0	5
1	1
2	2
3	14
4	8
5	16
6	7
7	12
8	18
9	15
10	11

图7-21 页表示意图

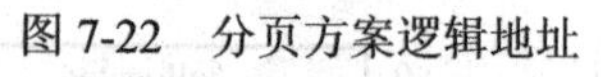

图7-22 分页方案逻辑地址

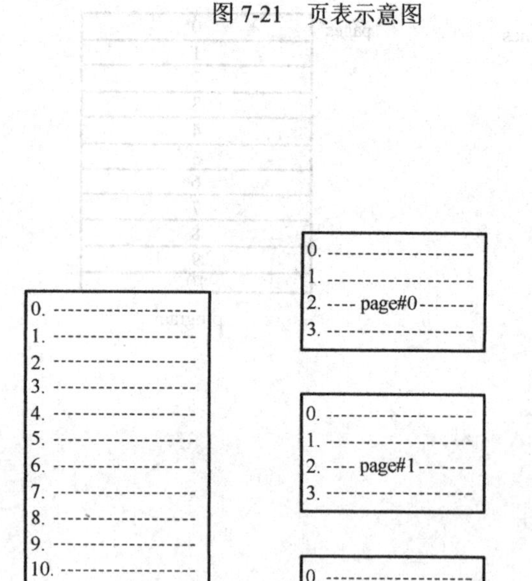

图7-23 进程逻辑地址空间

memory

frame#0: page#1 (0.–3.)
frame#1: 4.–7.
frame#2: page#2 (8.–11.)
frame#3: 12.–15.
frame#4: 16.–19.
frame#5: page#3 (20.–23.)

图7-24 页面存储图

现在尝试计算图7-23(a)中逻辑地址7对应的物理地址。根据上面的分页方案，该行地址被分在了1号页面的3号页内偏移上，也就是逻辑地址为<1, 3>。也可以这么理解：页面大小是4，所以1号页面的起始逻辑地址是4，加上页内偏移3，等于7。接着计算物理地址，1号页面装在0号页框中，该页框的物理起始地址是0，所以逻辑地址7对应的物理地址为0+3=

3。类似地，逻辑地址<0,3>对应的物理地址是 23。

我们得出分页方案物理地址的计算公式如下：

physical_address = frame_no*pagesize + offset

其中：

- frame_no 为页框号；
- pagesize 为页面大小；
- offset 为页内偏移。

在实际转换过程中，因为页面大小被规定是 2 的 *n* 次方，所以只需要将逻辑地址中的页号替换成页表中索引到的对应页框号，页内偏移保持不变，即可以生成对应的物理地址（如图 7-25 所示）。

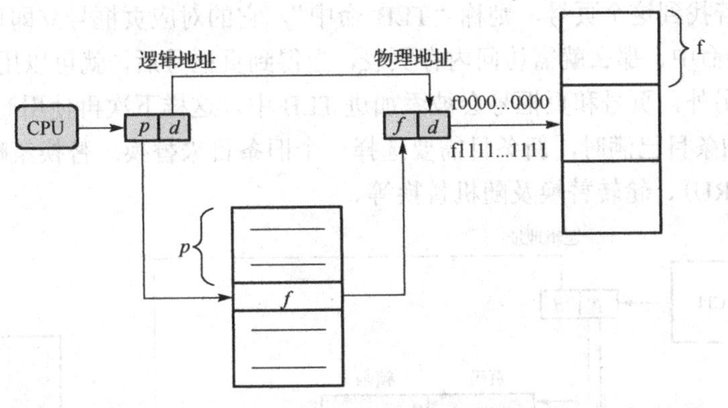

图 7-25　分页的硬件支持

分页逻辑地址如图 7-26 所示。逻辑地址的长度为 *m*，若页面大小为 2^n 字节，则低 *n* 位用于表示页内偏移，剩余的 *m–n* 位（高位）用于表示页号。根据计算机体系结构的不同，页面大小从 512B 到 1 GB 不等，在本书的实验环境中，页面大小为 4KB（如图 7-27 所示）。

Page No.	Page Offset
p	*d*
m − *n*	*n*

图 7-26　分页逻辑地址

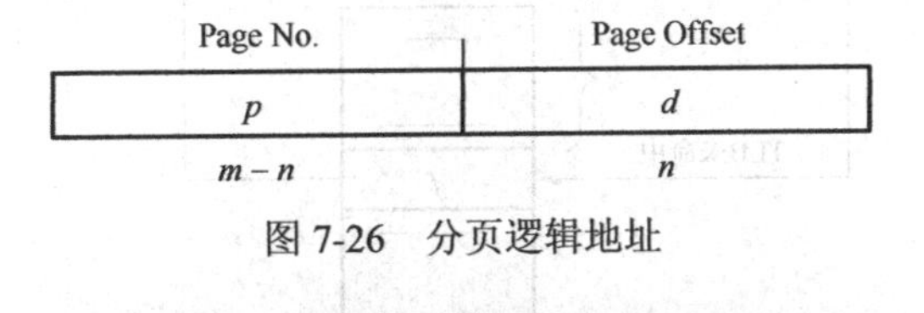

图 7-27　查看本书实验环境的页面大小

7.4.3　快表

本节专门讨论页表。页表的作用在于索引页面与页框的对应关系，每个进程都有一个页表，是进程的额外存储开销，页表由操作系统维护管理。每个操作系统都有自己保存页表的方法，页表的指针与其他寄存器（如 PC 寄存器）的值一起被存入进程控制块。当 CPU 需要启动一个进程时，它首先加载用户寄存器，并根据保存的页表来定义正确的硬件页表值。

如果页表比较小(如 256 个条目)，那么页表使用寄存器的效果还是令人满意的。但是，大多数现代计算机都允许非常大的页表(如 100 万个条目)。页表一般存放在内存中，将页表的内存起始地址放在页表基寄存器(Page-Table Base Register，PTBR)中，按照这种模式，每次访问一个内存地址需要访问两次内存：一次是为转换地址查询内存中的页表，另一次是访问转换后的内存地址。

这个问题的标准解决方案是采用专用的高速硬件缓冲区，称为相联存储器(Translation Look-aside Buffer，TLB)。TLB 条目由键(标签)和值组成，它的主要任务就是根据指定的键查找对应的值。现代的 TLB 查找硬件是指令流水线的一部分，基本上不添加任何性能负担。但 TLB 的容量比较小，只能存放少量的页表条目，因为 TLB 是被所有进程共享的，所以在 TLB 中保存的页表条目可能来自多个不同的进程。当 CPU 产生一个逻辑地址后，它的页号就被发送到 TLB 中，若找到这个页号，则称“TLB 命中”，它的对应页框号立即可用于组装物理地址；如果 TLB 未命中，那么就需访问内存页表，当得到页框号后，就可以用它来访问内存(如图 7-28 所示)。另外，页号和页框号会被添加进 TLB 中，这样下次再使用时就可以在 TLB 中命中。当 TLB 的条目已满时，新条目需要选择一个旧条目来替换。替换策略有很多，如最近最少使用替换(LRU)、轮转替换及随机替换等。

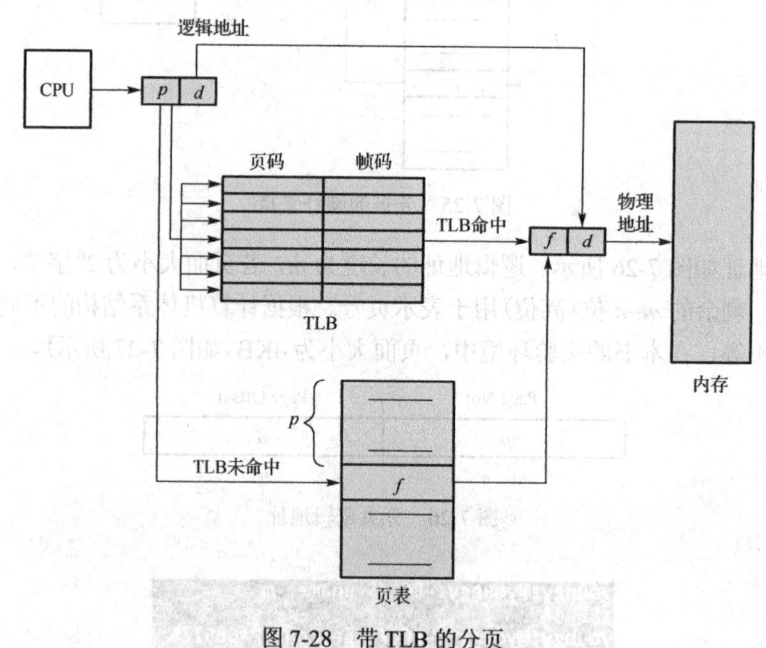

图 7-28　带 TLB 的分页

在 TLB 中查找到给定页号次数的百分比称为命中率(Hit Ratio)。80%的命中率意味着有 80%的时间可以在 TLB 中找到所需的页号。如果一次内存访问需要 100 ns，那么 TLB 命中时访问内存需要 80 ns。若 TLB 未命中，则要花费 2 倍的时间才能访问到内存中的所需字节，而有效内存访问时间(Effective Memory Access Time)，需要根据概率来进行加权：

$$\text{Effective Memory Access Time} = 0.80\times100 + 0.20\times200 = 120\ \text{ns}$$

也就是说，有效内存访问时间比单次访问内存多了 20%。假设命中率是 99%，则有效内存访问时间=0.99×100+0.01×200=101 ns，比单次访问内存只多了 1%的访问时间，可见 TLB 的命中率对访问效率的影响是非常大的。

7.4.4　多级页表

大多数现代计算机操作系统支持大逻辑地址空间。例如，假设某系统的逻辑地址是 32 位的，那么逻辑地址空间的大小是 2^{32} 字节；页面大小若是 4KB (2^{12})，则逻辑地址的页内偏移占 12 位，页号占 20 位；页表最大的条目数是 2^{20} 个，假设每个条目占 4 字节，则每个进程的页表大小最大是 2^{20}×4B=4MB，也就是要占用 1K 个连续的页框来存储它。连续分配的缺点是分配方案不灵活，为了解决这个问题，我们再次借用离散存储的思想，这回我们将页表再分页后离散存放，当然又会多出一个数据结构(页表目录，Page Table Directory)来记录页表页应对的页框，于是就形成了两级页表方案。

接上面这个例子，我们将逻辑地址的高 20 位拆成 2 个 10 位，p_1 记作页表页号，p_2 记作页号，页内偏移保持 12 位不变(如图 7-29 所示)。

Page No.		Page Offset
p_1	p_2	d
10	10	12

图 7-29　两级页表逻辑地址

图 7-30 展示了一个两级页表地址的转换过程，示例中的进程被分成了 8 个页面，所以它的页表共有 8 个条目，我们将该页表再分成 2 页：0～3 条目存入 7 号页框，4～7 条目存入 10 号页框，并把这个对应关系记录在页表目录中。在进行逻辑地址转换时，分 5 步。

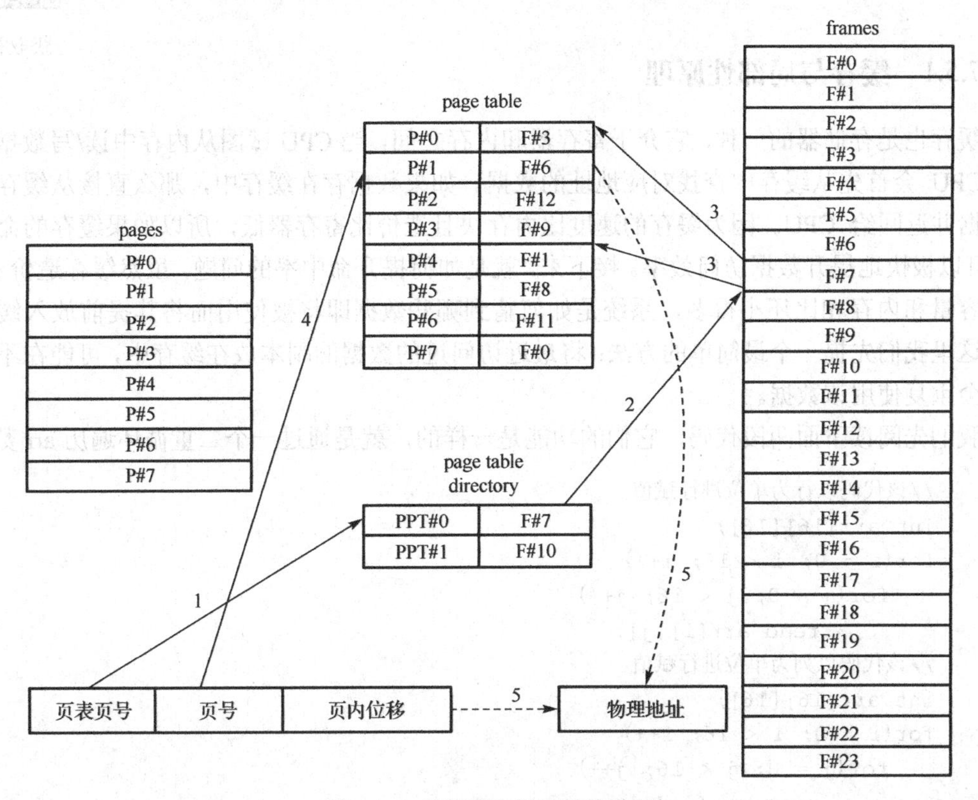

图 7-30　两级页表地址转换过程

(1)从逻辑地址中读出页表页号(Page of Page Table Number)，在页表目录中检索。

(2)得到该页表页在内存中的位置。

(3)读出页表页的内容。

(4)从逻辑地址中读出页号，在第(3)步读出的页表页中检索。

(5)得到页框号，再和页内偏移组合生成物理地址。

随着计算机技术的发展，目前早已进入64位系统时代，这也意味着逻辑地址的空间可达到2^{64}字节，这是一个巨大的数字，即使使用两级页表，进程页表页也将相当大。在这种情况下，我们可以如法炮制，将页表页继续分页得到三级页表，甚至是四级页表。在x86-64架构下Linux采用了四级页表方案，逻辑地址只使用了低48位，逻辑地址的分配如下。

(1)63～48位：保留未用；

(2)47～39位：页全局目录(Page Global Directory，PGD)；

(3)38～30位：页上级目录(Page Upper Directory，PUD)；

(4)29～21位：页中间目录(Page Middle Directory，PMD)；

(5)20～12位：页表项(Page Table Entry，PTE)；

(6)11～0位：页内偏移。

7.5 虚拟内存

获取视频

7.5.1 缓存与局部性原理

缓存也是存储器的一种，它介于寄存器和内存之间，当CPU试图从内存中读/写数据的时候，CPU会首先从缓存中查找对应地址的数据。如果数据存在缓存中，那么直接从缓存中读取数据并返回给CPU。因为缓存的速度比内存快且造价比寄存器低，所以如果缓存的命中率高，可以极快地提升数据访问效率。接下来，就是如何提升命中率的问题，虽然缓存造价不高，但其容量和内存相比还小得多，系统是如何猜到哪些数据即将被使用而将其提前放入缓存的呢？这里我们先提一个最简单的方法：将最近访问过的数据的副本放在缓存中，可能在不久的将来会重复使用该数据。

我们先阅读下面两段代码，它们的功能是一样的，就是通过一个二重循环遍历arr数组。

```
//该代码以行为单位进行赋值
int arr[16][16];
for(i = 0; i < 16; i++)
    for(j = 0; j < 16; j++)
        read arr[i][j];
//该代码以列为单位进行赋值
int arr[16][16];
for(i = 0; i < 16; i++)
    for(j = 0; j < 16; j++)
        read arr[j][i];
```

缓存的存储是以行为单位的，一个缓存一次会读入一行数据，如图7-31所示，每次循环都会将访问过的数组元素的副本存入缓存。假设这个缓存的每行都是 64 字节的，正好是 16 个 int 类型数据的长度，那么按第 1 段代码执行，在读取 arr[0][0]元素时，会将 arr[0][0]～a[0][15]整行的数据都加载进缓存(因为数组在内存中是按行的顺序连续存储的)，这样读取 arr[0][1]～[0][15]这 15 个元素的值时，没有必要再访问内存，直接可以从缓存中读取。若按第 2 段代码执行，缓存则会失去这个效果，因此第 1 段代码的执行效率是第 2 段代码的 8 倍。

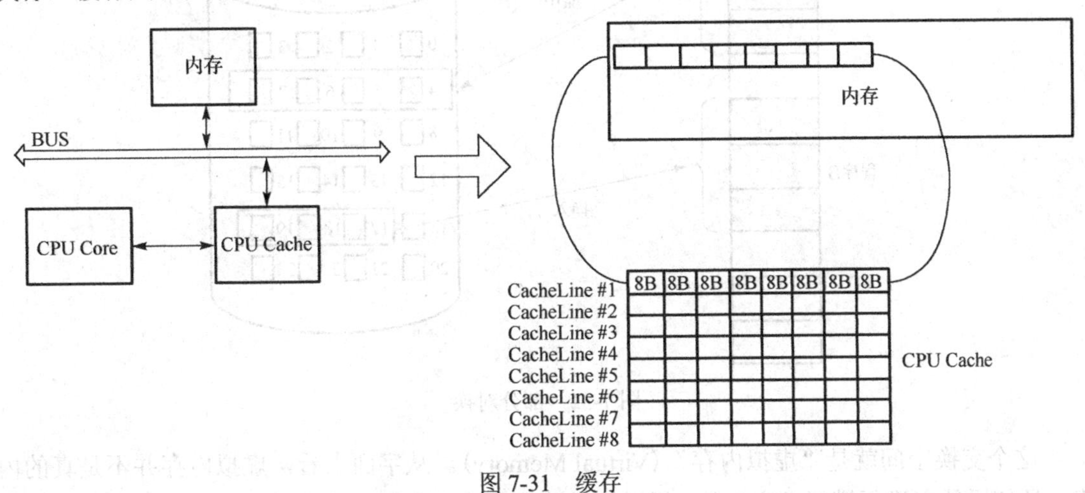

图 7-31 缓存

缓存的这个特性及刚才提到的方案都是基于程序局部性原理的，它是指 CPU 访问存储器时，无论采用的是读取指令还是存取数据，所访问的存储单元在一段时间内都趋向于一个较小的连续区域中。这包含两层意思：

(1) 空间局部性(Spatial Locality)：紧邻被访问单元的地方也将被访问。

(2) 时间局部性(Temporal Locality)：刚被访问的单元很快将再次被访问。

局部性原理与其说是原理，不如说是一种“现象”。这种局部性保证了层次化存储系统具有优秀的性能。

我们再回顾一下图 7-1，从金字塔顶端向下，容量越来越大，抽象一点来说就是，下级存储是上级存储的扩充。从下向上，速度越来越快，根据贪婪原则，我们总希望下一个访问的数据就在上层存储器中，于是上级存储器就成了下级存储器的缓存。按照这个思路，主存是缓存的扩充，缓存对主存进行了**缓存**(注：该词在这个地方理解为一个动作，即 Cache 的功能)。“外存是内存的扩充，内存是外存的缓存”这句话的确很绕口，希望读者学完本节后再体会这句话的含义。

7.5.2 虚拟内存

程序必须进入内存方可成为进程运行，内存空间是相当宝贵的，人们总希望更多的进程在内存中并发运行以提升 CPU 的利用率。为此，操作系统设计人员想出了一种方法，即进程运行时仅加载立即要运行的部分进入内存，而不必一次性全部加载，将其余部分暂时放在一个叫交换空间(Swap Space)的地方，这就是部分装入。交换空间通常是在外存(磁盘)上开辟的一块专用空间，由操作系统管理。部分对换是指系统可将内存中进程的部分换出内存至交换空间中，用以腾出内存空间，未来若要重新用到这部分，再从交换空间中将其对换回内存(如图 7-32 所示)。

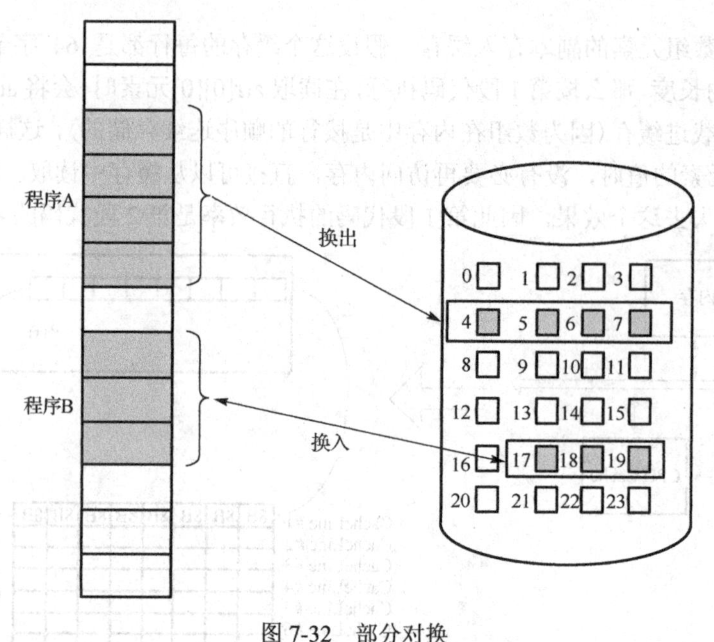

图 7-32　部分对换

这个交换空间就是“虚拟内存”(Virtual Memory)。从字面上看，虚拟内存并不是真的内存，是使用外存进行模拟而产生的，通过这种将逻辑内存和物理内存分开的方式，使得在有限物理内存的情况下为程序员提供巨大的虚拟内存。程序员不再需要考虑物理内存大小的限制，而只需关注如何解决问题。

本书以分页式内存管理方案为基础来讨论虚拟内存，进程在内存中的存储单位是页面，所以装入和对换的单位也是页面，下面我们讨论如何使用“按需调页”来实现虚拟内存——请求分页式虚拟内存管理方案。

7.5.3　请求调页

根据部分装入原则，进程可以只装入第一页进入内存，在运行过程中，需要哪一页再请求加载它，这个技术称为“请求调页”(Demand Paging)，常用于虚拟内存系统。对于请求调页的虚拟内存，页面只有在程序运行期间被请求时才能被加载。因此，从未访问的那些页面从不被加载到物理内存中。

使用这种方案需要一定的硬件支持，以区分内存的页面和磁盘的页面。在进程页表中要增加一个额外的“状态位”，当该位被设置为“有效”(Valid)时，说明该页面在内存中，当该位被设置为“无效”(Invalid)时，说明页面不在内存中而在交换空间中。如图 7-33 所示，A、C 及 F 三个页面在内存中，页表中对应的状态位“v”表示有效，若此时访问 B(1 号)页面，则会发现该页不在内存中而在交换空间中，此时会发出请求，系统将页面调入内存，这个请求称为“缺页中断”(Page Fault)。

缺页中断发出后会陷入操作系统代码，由操作系统处理这个中断，整个缺页中断处理过程如图 7-34 所示。

(1) 检查进程页表，确定页面状态是有效的还是无效的。

(2) 若页面状态无效，则说明尚未调入内存，于是发出缺页中断(陷阱)。

(3) 找到在交换空间(备用存储)中的页面，同时找到一个空闲的内存页框。

(4) 调用一个磁盘操作，将所需页面读到刚申请的页框中。

(5) 修改进程页表，修改状态位为“v”以指示该页面现在处于内存中。

(6) 缺页中断处理完毕后，返回进程被中断的地方继续运行。

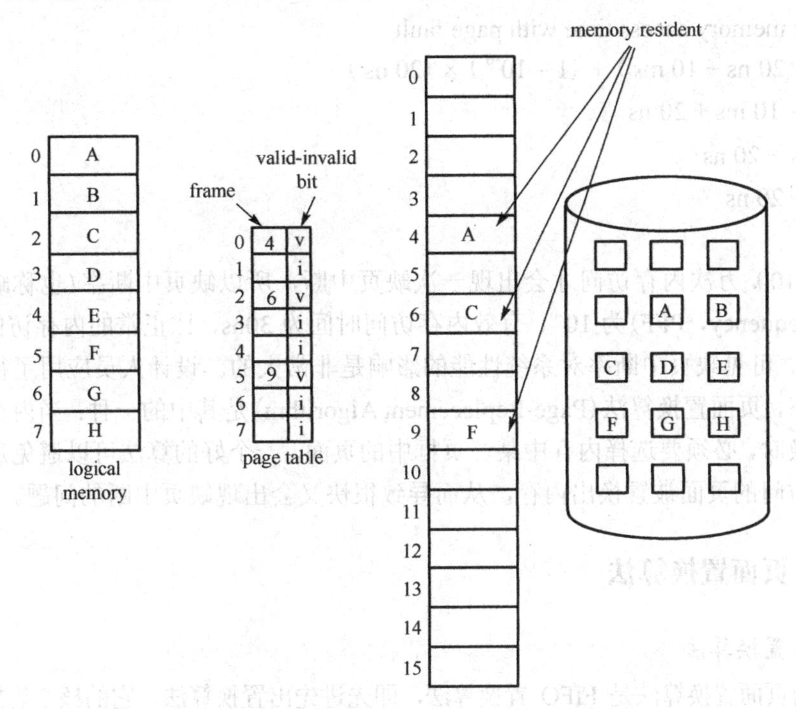

图 7-33　页面不存在内存中的页表

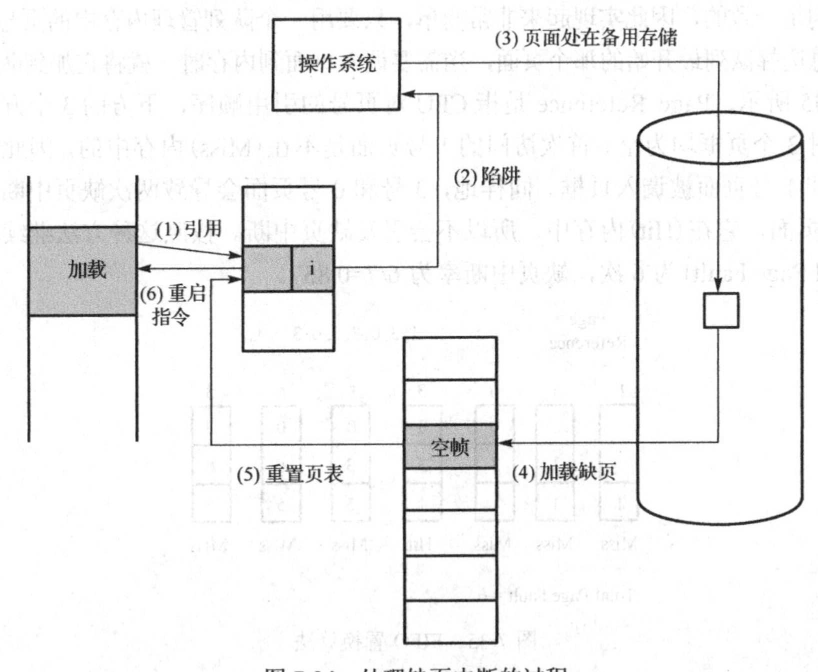

图 7-34　处理缺页中断的过程

请求调页虽然解决了内存无限大的问题，但是我们还要估算一下对系统性能的影响。处理一次缺页中断的时间不会太短，因为中断处理需要时间，而且磁盘的 I/O 操作速度较慢，大概需要 10ms。而内存的平均存取时间约为 20ns。如果假设每 100 万次内存访问才会出现一次缺页中断，那么有效内存访问时间

Effective memory access time with page fault

$= 10^{-6} \times (20\text{ ns} + 10\text{ ms}) + (1 - 10^{-6}) \times (20\text{ ns})$

$\approx 10^{-6} \times 10\text{ ms} + 20\text{ ns}$

$= 10^{-5}\text{ ms} + 20\text{ ns}$

$= 10\text{ ns} + 20\text{ ns}$

$= 30\text{ ns}$

因为每 100 万次内存访问才会出现一次缺页中断，所以缺页中断率(也称缺页错误率，Page-Fault Frequency，PFF)为 10^{-6}，有效内存访问时间为 30ns，比正常的内存访问时间(20ns)增加了 50%，可见缺页中断率对系统性能的影响是非常大的。设计人员应用了很多方案来降低缺页中断率，页面置换算法(Page-Replacement Algorithm)是其中的一种：当内存空间不足要进行页面置换时，必须要选择内存中某一页框中的页面，一个好的算法可以避免那些经常被访问或即将被访问的页面被置换出内存，从而导致很快又会出现缺页中断的问题。

7.5.4 页面置换算法

1. FIFO 置换算法

最简单的页面置换算法是 FIFO 置换算法，即先进先出置换算法。它的核心思想是，在必须置换页面时选择的是先进入内存的页面(也是在内存中待得最久的页面)，这个算法的思想和数据结构中的队列是一致的，因此实现起来非常简单，只要用一个队列管理内存中的页号即可，在进行置换时，总选择队列最开始的那个页面，当需要调入页面到内存时，就将它加到队列的尾部。

如图 7-35 所示，Page Reference 是指 CPU 对页号的引用顺序，下方的 3 个方框对应内存页框，初始时 3 个页框均为空。首次访问的 1 号页面是不在(Miss)内存中的，因此发生一次缺页中断，随即 1 号页面被调入页框。同样地，3 号和 0 号页面会导致两次缺页中断。第 4 次访问的是 3 号页面，它在(Hit)内存中，所以不会引发缺页中断。按照这种方法继续，最后总缺页次数(Total Page Fault)为 6 次，缺页中断率为 6/7=0.857。

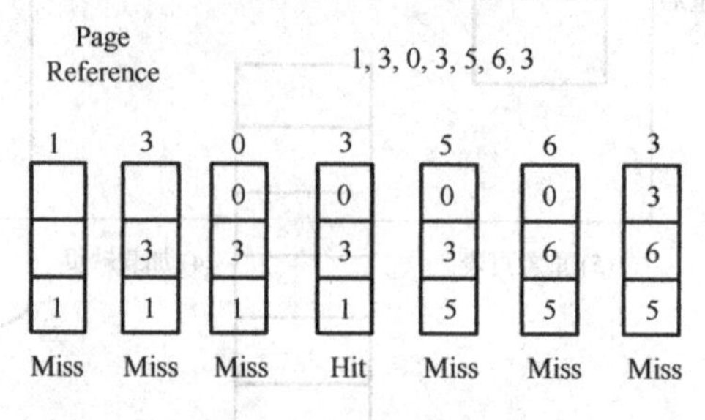

图 7-35 FIFO 置换算法

FIFO 置换算法易于理解和编程。然而，它的性能并不总是十分理想的。有一种情况：置

换的页面可能包含一个被大量使用的变量，它早就被初始化了，但仍被不断使用，选择它进行置换是不明智的。

当然，如果我们增加进程所得到的页框数量，按道理缺页中断次数应该下降，但也存在例外情况，称为“Belady 异常”：随着分配页框数量的增加，FIFO 置换算法的缺页中断率可能会增加。比如，页面引用序列：3, 2, 1, 0, 3, 2, 4, 3, 2, 1, 0, 4，若分配其 3 个页框，则产生 9 次缺页中断，若分配其 4 个页框，缺页中断会增加到 10 次。

2. 最优页面置换算法

最优页面置换(Optimal Page-replacement，OPT)算法具有所有算法中最低的缺页中断率，并且不会发生 Belady 异常。这种算法总是置换最长时间不会使用的页面，它能确保对于给定数量的帧会产生最低的缺页中断率。

例如，如图 7-36 所示，针对页面引用序列：7, 0, 1, 2, 0, 3, 0, 4, 2, 3, 0, 3, 2, 3，页框数量为 4，初始为空，按 OPT 算法会产生 6 次缺页中断。前 4 个引用会产生缺页中断，以填满 4 个空闲页框。在第 6 次访问 3 号页面时，因为 7 号页面在未来都不会被使用，因此被选中置换，同样的原因当访问 4 号页面时，会选择之后都不再被用到的 1 号页面进行置换。

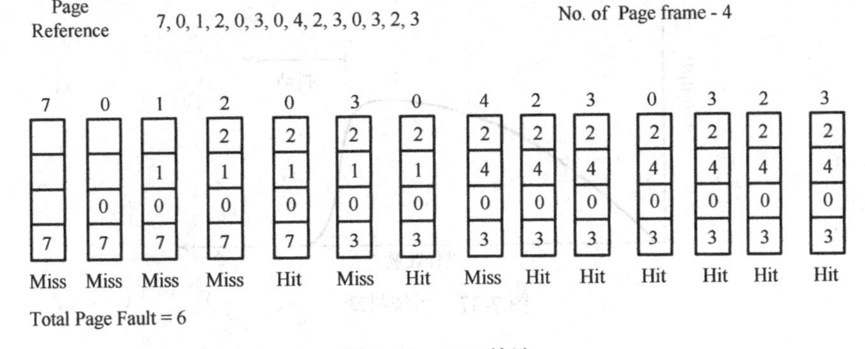

图 7-36　OPT 算法

OPT 算法是最优的，但可惜的是它并不能实现，因为我们无法预测未来的页面访问序列，OPT 算法的作用在于评估其他转换算法的好坏，例如，如果知道一个算法不是最优的，但是最坏情况较最优情况相差不大于 12.3%，平均情况相差不大于 4.7%，那么该算法也是很有用的。

3. LRU 置换算法

OPT 算法虽然不可行，但是我们可以设计它的近似算法。虽然没有办法预测未来，但是我们可以观察过去，用过去发生的事件来近似预测未来，如果在过去的一段时间内，有一个页面是最长时间未用过的页面，那么我们也认为它在未来用到的机会比较少，于是就挑选它作为置换对象。这种方法称为最近最少使用(Least-Recent-Used，LRU)算法。

我们仍然使用图 7-36 所示的例子，在访问 3 号页面时，因为 7 号页面在过去的时间内最久未被使用，所以选择它进行置换，同样的道理，在访问 4 号页面时选择 1 号页面进行置换。最后 LRU 算法的缺页中断次数和 OPT 算法相同，都是 6 次。

当然 LRU 算法只是 OPT 算法的近似算法，它的效果并不是每次都和 OPT 算法相同，在有些访问顺序和页框数量前提下，其效果差很多。

7.5.5 系统抖动

1. 抖动原因

系统抖动(Thrashing)是指因被调出的页面又立刻被调入而形成的频繁调入调出现象。虚拟存储管理中可能出现系统抖动的现象，即刚被淘汰(从内存调到外存中)的页面，时隔不久又要被访问，因而又要把它调入，调入不久又再次被淘汰，如此反复，使得整个系统页面的调入调出工作非常频繁，以致CPU大部分时间都用在来回进行页面调度上，只有很少一部分时间用于进程的实际计算。系统抖动也称为颠簸。

操作系统监视CPU利用率。如果CPU利用率太低，那么可以通过向系统引入新的进程来增加多道程度，CPU利用率也会随之提升(如图7-37所示)。随着进程并发数量的增加，所需要的内存也增大，慢慢开始出现缺页中断，进程之间开始抢夺页框，如果进程数量持续增加，进程大部分的时间都用来请求调页，系统性能会急剧下降，甚至趋于崩溃，系统应立即采取措施加以排除，如关闭若干进程降低并发数量。

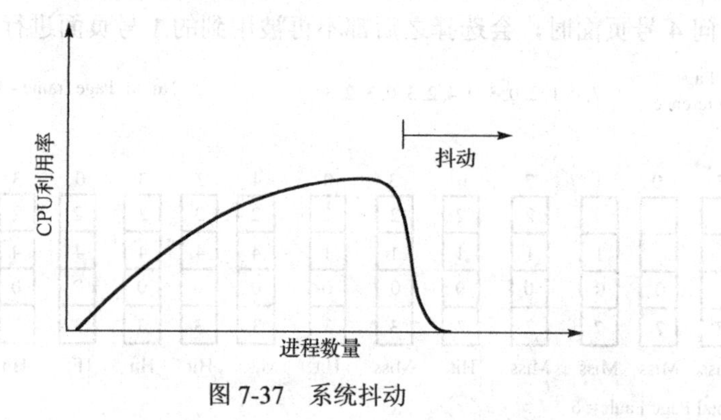

图7-37 系统抖动

2. 工作集模型

为了避免发生系统抖动现象，操作系统基于局部性原理设计了工作集(Working Set)模型。工作集(或驻留集)是指在某段时间间隔内，进程要访问的页面集合。经常被使用的页面需要在工作集中，而长期不使用的页面要从工作集中被丢弃。为了防止出现系统抖动现象，需要选择合适的工作集大小。

这个模型采用参数Δ定义工作集窗口(Working Set Window)：检查某时刻最近Δ个页面引用，这最近Δ个页面引用的页面集合称为工作集。如图7-38所示，Δ=10，t_1时刻的工作集WS(t_1)={1,2,5,6,7}，到t_2时，WS(t_2)={3,4}。

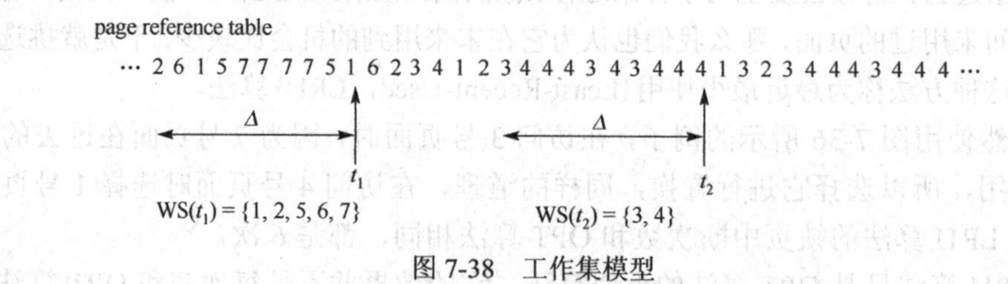

图7-38 工作集模型

工作集的精度取决于Δ的选择。如果Δ太小，那么它不能包含整个局部；如果Δ太大，那么它可能包含多个局部。一旦选中了Δ，工作集模型的使用就很简单了。让操作系统跟踪每个进程的工作集，并为进程分配大于其工作集的页框。若还有空闲页框，则可以再调一个进程到内存中，以增加并发进程数量。如果所有工作集之和超过了可用页框的总数，那么操作系统会暂停一个进程，将其页面调出且将其页框分配给其他进程，防止出现系统抖动现象。

3. 缺页中断率

防止系统抖动的又一种策略是监测系统的缺页中断率。由于系统抖动会产生高缺页中断率，因此可以通过监测缺页中断率来判断是否发生系统抖动。如图7-39所示，一开始对某进程增加帧数(页框数)，其缺页中断率会快速下降，在帧数增加到一定程度后再增加页框数量，缺页中断率的下降幅度变得缓慢。我们只需要将缺页中断率控制在上下限范围内即可，若低于下限，则回收进程页框，若高于上限，则为进程再分配页框，以此对缺页中断率进行控制，防止系统抖动。

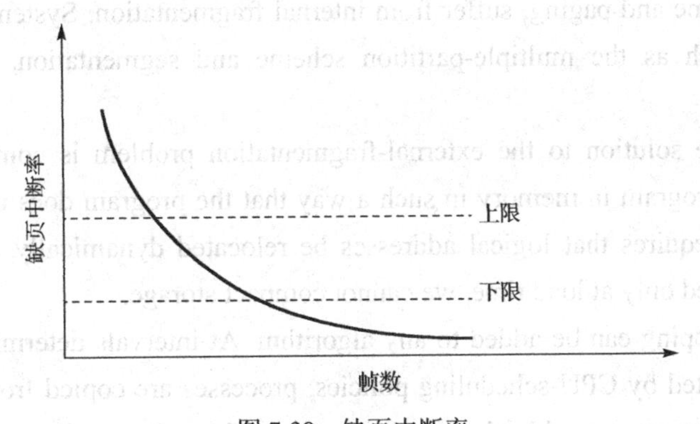

图7-39 缺页中断率

7.6 Reading Materials

7.6.1 Overview

Memory-management algorithms for multiprogrammed operating systems range from the simple single-user system approach to segmentation and paging. The most important determinant of the method used in a particular system is the hardware provided. Every memory address generated by the CPU must be checked for legality and possibly mapped to a physical address. The checking cannot be implemented (efficiently) in software. Hence, we are constrained by the hardware available.

The various memory-management algorithms (contiguous allocation, paging, segmentation, and combinations of paging and segmentation) differ in many aspects. In comparing different memory-management strategies, we use the following considerations:

Hardware support. A simple base register or a base-limit register pair is sufficient for the single-and multiple-partition schemes, whereas paging and segmentation need mapping tables to define the address map.

Performance. As the memory-management algorithm becomes more complex, the time required to map a logical address to a physical address increases. For the simple systems, we need only compare or add to the logical address, operations that are fast. Paging and segmentation can be as fast if the mapping table is implemented in fast registers. If the table is in memory, however, user memory accesses can be degraded substantially. A TLB can reduce the performance degradation to an acceptable level.

Fragmentation. A multiprogrammed system will generally perform more efficiently if it has a higher level of multiprogramming. For a given set of processes, we can increase the multiprogramming level only by packing more processes into memory. To accomplish this task, we must reduce memory waste, or fragmentation. Systems with fixed-sized allocation units, such as the single-partition scheme and paging, suffer from internal fragmentation. Systems with variable-sized allocation units, such as the multiple-partition scheme and segmentation, suffer from external fragmentation.

Relocation. One solution to the external-fragmentation problem is compaction. Compaction involves shifting a program in memory in such a way that the program does not notice the change. This consideration requires that logical addresses be relocated dynamically, at execution time. If addresses are relocated only at load time, we cannot compact storage.

Swapping. Swapping can be added to any algorithm. At intervals determined by the operating system, usually dictated by CPU-scheduling policies, processes are copied from main memory to a backing store and later are copied back to main memory. This scheme allows more processes to be run than can be fit into memory at one time. In general, PC operating systems support paging, and operating systems for mobile devices do not.

Sharing. Another means of increasing the multiprogramming level is to share code and data among different processes. Sharing generally requires that either paging or segmentation be used to provide small packets of information (pages or segments) that can be shared. Sharing is a means of running many processes with a limited amount of memory, but shared programs and data must be designed carefully.

Protection. If paging or segmentation is provided, different sections of a user program can be declared execute-only, read-only, or read-write. This restriction is necessary with shared code or data and is generally useful in any case to provide simple run-time checks for common programming errors.

It is desirable to be able to execute a process whose logical address space is larger than the available physical address space. Virtual memory is a technique that enables us to map a large logical address space onto a smaller physical memory. Virtual memory allows us to run extremely large processes and to raise the degree of multiprogramming, increasing CPU utilization. Further, it frees application programmers from worrying about memory availability. In addition, with virtual memory, several processes can share system libraries and memory. With virtual memory, we can also use an efficient type of process creation known as copy-on-write, wherein parent and child processes share actual pages of memory.

Virtual memory is commonly implemented by demand paging. Pure demand paging never brings in a page until that page is referenced. The first reference causes a page fault to the operating system. The operating-system kernel consults an internal table to determine where the page is located on the backing store. It then finds a free frame and reads the page in from the backing store. The page table is updated to reflect this change, and the instruction that caused the page fault is restarted. This approach allows a process to run even though its entire memory image is not in main memory at once. As long as the page-fault rate is reasonably low, performance is acceptable.

We can use demand paging to reduce the number of frames allocated to a process. This arrangement can increase the degree of multiprogramming (allowing more processes to be available for execution at one time) and—in theory, at least—the CPU utilization of the system. It also allows processes to be run even though their memory requirements exceed the total available physical memory. Such processes run in virtual memory.

If total memory requirements exceed the capacity of physical memory, then it may be necessary to replace pages from memory to free frames for new pages. Various page-replacement algorithms are used. FIFO page replacement is easy to program but suffers from Belady's anomaly. Optimal page replacement requires future knowledge. LRU replacement is an approximation of optimal page replacement, but even it may be difficult to implement. Most page-replacement algorithms, such as the second-chance algorithm, are approximations of LRU replacement.

In addition to a page-replacement algorithm, a frame-allocation policy is needed. Allocation can be fixed, suggesting local page replacement, or dynamic, suggesting global replacement. The working-set model assumes that processes execute in localities. The working set is the set of pages in the current locality. Accordingly, each process should be allocated enough frames for its current working set. If a process does not have enough memory for its working set, it will thrash. Providing enough frames to each process to avoid thrashing may require process swapping and scheduling.

Most operating systems provide features for memory mapping files, thus allowing file I/O to be treated as routine memory access. The Win32 API implements shared memory through memory mapping of files.

Kernel processes typically require memory to be allocated using pages that are physically contiguous. The buddy system allocates memory to kernel processes in units sized according to a power of 2, which often results in fragmentation. Slab allocators assign kernel data structures to caches associated with slabs, which are made up of one or more physically contiguous pages. With slab allocation, no memory is wasted due to fragmentation, and memory requests can be satisfied quickly.

In addition to requiring us to solve the major problems of page replacement and frame allocation, the proper design of a paging system requires that we consider prepaging, page size, TLB reach, inverted page tables, program structure, I/O interlock and page locking, and other issues.

7.6.2 Virtual Memory

Virtual memory combines active RAM and inactive memory on DASD to form a large range of contiguous addresses.

In computing, virtual memory, or virtual storage is a memory management technique that provides an "idealized abstraction of the storage resources that are actually available on a given machine" which "creates the illusion to users of a very large (main) memory".

The computer's operating system, using a combination of hardware and software, maps memory addresses used by a program, called virtual addresses, into physical addresses in computer memory. Main storage, as seen by a process or task, appears as a contiguous address space or collection of contiguous segments. The operating system manages virtual address spaces and the assignment of real memory to virtual memory. Address translation hardware in the CPU, often referred to as a memory management unit (MMU), automatically translates virtual addresses to physical addresses. Software within the operating system may extend these capabilities to provide a virtual address space that can exceed the capacity of real memory and thus reference more memory than is physically present in the computer.

The primary benefits of virtual memory include freeing applications from having to manage a shared memory space, ability to share memory used by libraries between processes, increased security due to memory isolation, and being able to conceptually use more memory than might be physically available, using the technique of paging or segmentation.

7.6.3 Segmented Virtual Memory

Some systems, such as the Burroughs B5500, use segmentation instead of paging, dividing virtual address spaces into variable-length segments. A virtual address here consists of a segment number and an offset within the segment. The Intel 80286 supports a similar segmentation scheme as an option, but it is rarely used. Segmentation and paging can be used together by dividing each segment into pages; systems with this memory structure, such as Multics and IBM System/38, are usually paging-predominant, segmentation providing memory protection.

In the Intel 80386 and later IA-32 processors, the segments reside in a 32-bit linear, paged address space. Segments can be moved in and out of that space; pages there can "page" in and out of main memory, providing two levels of virtual memory; few if any operating systems do so, instead using only paging. Early non-hardware-assisted x86 virtualization solutions combined paging and segmentation because x86 paging offers only two protection domains whereas a VMM / guest OS / guest applications stack needs three. :22 The difference between paging and segmentation systems is not only about memory division; segmentation is visible to user processes, as part of memory model semantics. Hence, instead of memory that looks like a single large space, it is structured into multiple spaces.

This difference has important consequences; a segment is not a page with variable length or a simple way to lengthen the address space. Segmentation that can provide a single-level memory model in which there is no differentiation between process memory and file system consists of only a list of segments（files）mapped into the process's potential address space.

This is not the same as the mechanisms provided by calls such as mmap and Win32's MapViewOfFile, because inter-file pointers do not work when mapping files into semi-arbitrary places. In Multics, a file（or a segment from a multi-segment file）is mapped into a segment in the address space, so files are always mapped at a segment boundary. A file's linkage section can contain pointers for which an attempt to load the pointer into a register or make an indirect reference through it causes a trap. The unresolved pointer contains an indication of the name of the segment to which the pointer refers and an offset within the segment; the handler for the trap maps the segment into the address space, puts the segment number into the pointer, changes the tag field in the pointer so that it no longer causes a trap, and returns to the code where the trap occurred, re-executing the instruction that caused the trap. This eliminates the need for a linker completely and works when different processes map the same file into different places in their private address spaces.

7.7 实验6 进程内存空间

获取视频

1. 实验目的

掌握查看程序目标代码逻辑地址的技能，掌握观察程序运行后各部分内存地址的技巧，深入理解逻辑地址和物理地址的意义和关联。

2. 实验方法

按实验过程完成实验内容，观察实验结果，可以跟随实验视频完成。

3. 实验内容

1) 逻辑地址和物理地址

(1) 先编写一段完整而简单的代码。

```
//sample.c
int sum(int x, int y){
 return x+y;
}

int main(){
    sum(2,3);
    return 0;
}
```

(2) 使用gcc进行编译和汇编，但不链接成可执行文件，得到目标代码sample.o。

```
$ gcc -g -c sample.c
```

(3) 使用 objdump 对 sample.o 进行反编译，查看其汇编代码。

```
$ objdump -d sample.o
```

在本书的 Deepin15.11 环境下，显示结果如下。

```
sample.o:     文件格式 elf64-x86-64

Disassembly of section .text:

0000000000000000 <sum>:
int sum(int x, int y){
   0:  55                    push   %rbp
   1:  48 89 e5              mov    %rsp,%rbp
   4:  89 7d fc              mov    %edi,-0x4(%rbp)
   7:  89 75 f8              mov    %esi,-0x8(%rbp)
    return x+y;
   a:  8b 55 fc              mov    -0x4(%rbp),%edx
   d:  8b 45 f8              mov    -0x8(%rbp),%eax
  10:  01 d0                 add    %edx,%eax
}
  12:  5d                    pop    %rbp
  13:  c3                    retq

0000000000000014 <main>:
int main(){
  14:  55                    push   %rbp
  15:  48 89 e5              mov    %rsp,%rbp
   sum(2,3);
  18:  be 03 00 00 00        mov    $0x3,%esi
  1d:  bf 02 00 00 00        mov    $0x2,%edi
  22:  e8 00 00 00 00        callq  27 <main+0x13>
   return 0;
  27:  b8 00 00 00 00        mov    $0x0,%eax
  2c:  5d                    pop    %rbp
  2d:  c3                    retq
```

(4) 请特别关注上述结果中加粗的部分，从 0 开始递增，直到 2d 为止。这些数字即程序的逻辑地址，这种地址也称虚拟地址，对每个程序而言，它总是从 0 开始的，每个地址对应一条指令，我们把从"0:"这行开始的 4 行输出单独拿出来分析一下[所有的数字都是以十六进制数(即 4 个二进制数)编写的]：

```
0: 55                    push   %rbp
1: 48 89 e5              mov    %rsp,%rbp
4: 89 7d fc              mov    %edi,-0x4(%rbp)
7: 89 75 f8              mov    %esi,-0x8(%rbp)
```

- 0 是每个程序的起始逻辑地址；
- 0x55 是 push %rbp 汇编语句对应的机器指令， 正好 1 字节，因此第 2 条指令的逻辑地址=第 1 条指令逻辑地址+1=1；

- 0x48 89 e5 这 3 字节的机器指令对应的汇编语句是 mov %rsp, %rbp，本书不去解释这些汇编语句的作用，只需观察逻辑地址的递增规律，因为这条指令占了 3 字节，所以下一条指令的逻辑地址=本条指令逻辑地址+3=4；
- 后面的规律请读者自行分析。

(5) 逻辑地址 0x0～0x13 对应的是 sum()函数的所有指令，0x14～0x2d 对应的是 main()函数的所有指令，在逻辑地址 0x22 处执行了 call 指令，要调用 sum()函数，但此处并没有指明 sum()的逻辑地址，因为这段目标代码还未链接，所以很多地址有可能还会发生变化。我们去掉-g -c 参数，用 gcc 对 sample.c 执行编译、链接，生成可执行文件 sample，再用 objdump 对其进行反编译。

```
$ gcc sample.c -o sample
$ objdump -d sample
#输出的部分结果如下
0000000000000660 <sum>:
 660:   55                      push   %rbp
 661:   48 89 e5                mov    %rsp,%rbp
 664:   89 7d fc                mov    %edi,-0x4(%rbp)
 667:   89 75 f8                mov    %esi,-0x8(%rbp)
 66a:   8b 55 fc                mov    -0x4(%rbp),%edx
 66d:   8b 45 f8                mov    -0x8(%rbp),%eax
 670:   01 d0                   add    %edx,%eax
 672:   5d                      pop    %rbp
 673:   c3                      retq

0000000000000674 <main>:
 674:   55                      push   %rbp
 675:   48 89 e5                mov    %rsp,%rbp
 678:   be 03 00 00 00          mov    $0x3,%esi
 67d:   bf 02 00 00 00          mov    $0x2,%edi
 682:   e8 d9 ff ff ff          callq   660 <sum>
 687:   b8 00 00 00 00          mov    $0x0,%eax
 68c:   5d                      pop    %rbp
 68d:   c3                      retq
 68e:   66 90                   xchg   %ax,%ax
```

不难发现，sum()的起始逻辑地址变为了 0x660，main()的起始逻辑地址变成了 0x674，在 0x682 处，我们找到 call 660<sum>指令，旨在调用 sum()函数，即跳转到 sum()的起始地址处。

(6) 物理地址是内存单元中的地址，是指令和数据真实的内存地址，而逻辑地址是面向程序而言的。CPU 在执行指令的时候(如上面提到的 call 660)，会先将逻辑地址经一个 MMU 硬件设备转换成物理地址，再进行跳转。

2) 进程的内存映像

(1) 以 32 位机器为例，地址总线是 32 位的，可寻址的最大内存空间是 2^{32} 字节，即 4GB。每个运行的进程都可以获得一个 4GB 的逻辑地址空间，这个空间被分成两部分：内核空间和用户空间，其中用户空间为从0x00000000到0xC0000000，共3GB，而内核空间为从0xC0000000到0xFFFFFFFF的1GB空间（如图7-40所示）。

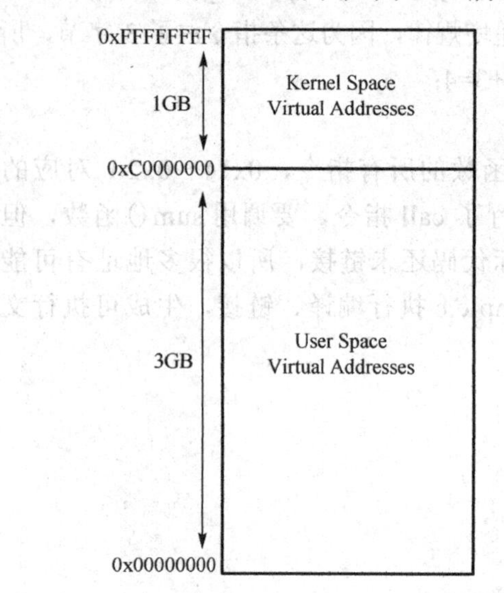

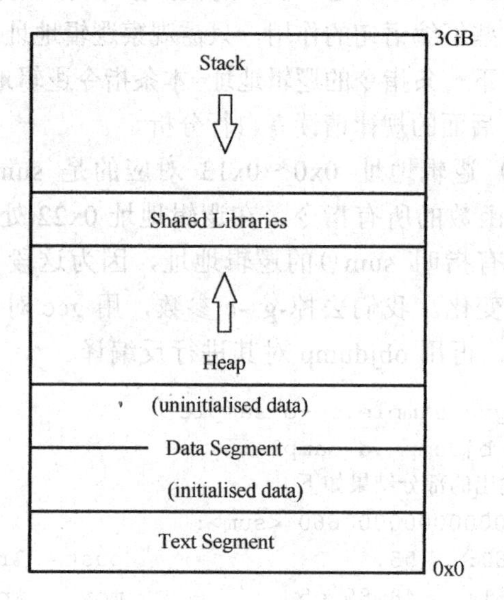

图 7-40 用户空间和内核空间

(2)用户进程的虚拟空间内容就是我们之前学过的进程内存映像，它共包括四部分内容：

- 代码段：是只读的，存放要执行的指令；
- 数据段：存放全局或静态变量；
- 堆：运行时分配的内存[如用malloc()函数申请的内存]；
- 栈：存放局部变量和函数返回地址。

因为堆、栈段的内容要程序运行后才能观察到，我们先看一下数据段的情况，再编写一个sample2.c，代码如下：

```
//sample2.c
int global_var = 5;

int main(){
 static int static_var = 6;
    return 0;
}
```

然后进行以下操作：

```
$ gcc -g -c sample2.c
$ objdump -s -d sample2
#部分结果如下
sample.o:     文件格式 elf64-x86-64

Contents of section.text:
 0000 554889e5 b8000000 005dc3           UH.....].
Contents of section.data:
 0000 05000000 06000000                  ........
```

```
...
...
...
Disassembly of section.text:

0000000000000000 <main>:
   0:    55                      push   %rbp
   1:    48 89 e5                mov    %rsp,%rbp
   4:    b8 00 00 00 00          mov    $0x0,%eax
   9:    5d                      pop    %rbp
   a:    c3                      retq
```

section.data 是数据段，里面已经写好了 global_var 和 static_var 的初始值；再看一看 section.text 中的十六进制数和下面 main()函数中的十六进制数，是不是很匹配？上面的就是 main()函数指令集合在代码段中的存放形式，下面的是 objdump 为了方便我们阅读而对其重新排版后的形式。

(3) 向 sample2.c 中再添加一个局部变量，并使用 malloc()在运行时申请一段内存空间，另存为 sample3.c，代码如下：

```
//sample3.c
#include <stdlib.h>
#include <unistd.h>
#include <stdio.h>

int global_var = 5;

int main(){
 static int static_var = 6;
    int local_var = 7;
    int*p = (int*)malloc(100);

 //Be careful: we must use %lx to show a 64bits address!!
    printf("the global_var address is %lx\n", &global_var);
    printf("the static_var address is %lx\n", &static_var);
    printf("the local_var address is %lx\n", &local_var);
    printf("the address which the p points to%lx\n", p);

    free(p);
    sleep(1000);//We need watch the process state, so let it sleep deeply.
    return 0;
 }
```

然后对其进行正常编译、链接，再运行它，它很快会进入 sleep 状态。

```
$ gcc sample3.c -o sample3
$ ./sample3
#输出结果如下
the global_var address is 55b3c179d048
```

```
the static_var address is 55b3c179d04c
the local_var address is 7ffdab0f3704
the address which the p points to55b3c2937010
```

(4)重新打开一个终端窗口，执行如下指令：

```
$ ps -e
 18288 pts/2    00:00:00 sample3
 18294 pts/1    00:00:00 ps
##我们把 sample3 的 pid：18288 记下来，每个机器的 pid 都不一样，请注意这一点！
$ cat /proc/18288/maps
##输出结果如下
55b3c159c000-55b3c159d000        r-xp       00000000        08:01        833578
/home/youngyt/Desktop/L12.MemoryManagement/sample3
55b3c179c000-55b3c179d000        r--p       00000000        08:01        833578
/home/youngyt/Desktop/L12.MemoryManagement/sample3
55b3c179d000-55b3c179e000        rw-p       00001000        08:01        833578
/home/youngyt/Desktop/L12.MemoryManagement/sample3
55b3c2937000-55b3c2958000 rw-p 00000000 00:00 0      [heap]
7f2b0f9c8000-7f2b0fb5d000        r-xp       00000000        08:01        922700
/usr/lib/x86_64-linux-gnu/libc-2.24.so
7f2b0fb5d000-7f2b0fd5d000        ---p       00195000        08:01        922700
/usr/lib/x86_64-linux-gnu/libc-2.24.so
7f2b0fd5d000-7f2b0fd61000        r--p       00195000        08:01        922700
/usr/lib/x86_64-linux-gnu/libc-2.24.so
7f2b0fd61000-7f2b0fd63000        rw-p       00199000        08:01        922700
/usr/lib/x86_64-linux-gnu/libc-2.24.so
7f2b0fd63000-7f2b0fd67000 rw-p 00000000 00:00 0
7f2b0fd67000-7f2b0fd8a000        r-xp       00000000        08:01        922248
/usr/lib/x86_64-linux-gnu/ld-2.24.so
7f2b0ff69000-7f2b0ff6b000 rw-p 00000000 00:00 0
7f2b0ff87000-7f2b0ff8a000 rw-p 00000000 00:00 0
7f2b0ff8a000-7f2b0ff8b000        r--p       00023000        08:01        922248
/usr/lib/x86_64-linux-gnu/ld-2.24.so
7f2b0ff8b000-7f2b0ff8c000        rw-p       00024000        08:01        922248
/usr/lib/x86_64-linux-gnu/ld-2.24.so
7f2b0ff8c000-7f2b0ff8d000 rw-p 00000000 00:00 0
7ffdab0d5000-7ffdab0f7000 rw-p 00000000 00:00 0       [stack]
7ffdab13b000-7ffdab13e000 r--p 00000000 00:00 0       [vvar]
7ffdab13e000-7ffdab140000 r-xp 00000000 00:00 0             [vdso]
ffffffffff600000-ffffffffff601000 r-xp 00000000 00:00 0       [vsyscall]
```


输出结果包含6列，每列的标题分别是 Address, Permissions, Offset, Device, inode, Pathname。

- Address: The start address and the end address in the process that the mapping occupies.
- Permissions: Readable, writable, executable, private, shared.
- Offset: The offset in the file where the mapping begins. Not every mapping is mapped from a file, so the value of offset is zero.
- Device: In the form “major number : minor number” of the associated device, as in, the file

from which this mapping is mapped from. Again, the mappings which are not mapped from any file, value is 00:00.

- inode: Inode of the related file.
- Pathname: The path of the related file. Its blank in case there is no related file.

(5) 现在找到这个进程在内存中四部分所在的地址区间：

代码段：第_______行

数据段：第_______行

堆：第_______行

栈：第_______行

第8章 外存管理

8.1 磁盘结构

获取视频

8.1.1 硬件结构

磁盘或硬盘(Magnetic Disk 或 Hard disk)为现代操作系统提供后备存储。图 8-1 是磁盘硬件构造图，磁盘的主体是一组圆形的盘片，盘片的两面都涂有磁性材料，这些磁性材料可以保存二进制信息，这也是磁盘名称的来历。所有的盘片会绕着主轴旋转，普通磁盘驱动器的转速包括 5400、7200、10000 及 15000RPM(Rotation Per Minute，每分钟转数)。在每个盘面上面都悬浮着一个读/写磁头，磁头附着在传动臂(Disk Arm，或称磁臂上)，传动臂可以左右摆动，摆的范围在盘片的半径内，它的摆动带动所有磁头一起移动。磁盘驱动器通过称为 I/O 总线(I/O Bus)的一组电缆连到计算机上。有多种可用总线接口，包括磁盘接口(Advanced Technology Attachment，ATA)、串行 ATA (Serial ATA，SATA)、外部串行 ATA(External Serial ATA，ESATA)、通用串口总线(Universal Serial Bus，USB)及光纤通道(Fiber Channel，FC)。

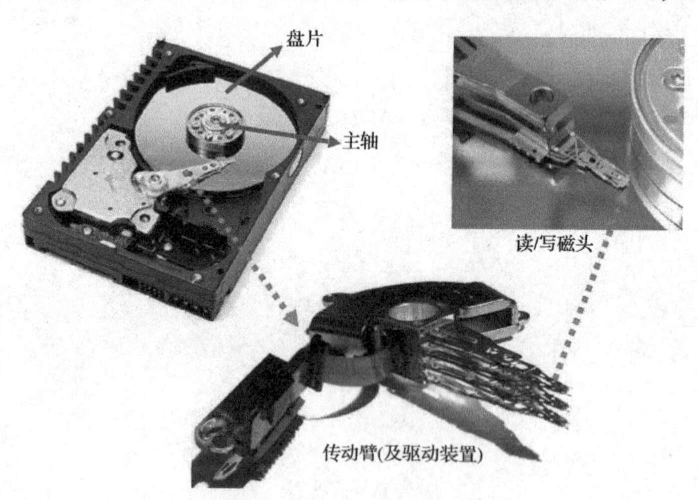

图 8-1　磁盘硬件构造图

盘片是数据的载体，它的存储结构是规则的圆形，它的表面被逻辑地分成若干同心圆，称为磁道(Track)，如图 8-2 所示，将每个磁道划分成相同多的片段，称为扇区(Sector)。所有盘片同一磁头下的磁道集合形成了柱面(Cylinder)，最外面的柱面是 0 号柱面，再按照从外到内的顺序对柱面进行编号。每个磁盘驱动器有数千个同心柱面，而每个磁道可能包括数百个扇区。扇区在有些书中也叫物理块(Block)。通常扇区的大小是 512 字节，现在整个磁盘的存储容量大多以 TB 来计算。

从图 8-2 中不难看出，每个磁道的扇区数量相同，所以内圈扇区存储密度要高于外圈扇区，因为扇区的大小是固定的，所以盘片在旋转时转速是恒定的，磁头无论是在内圈还是在外圈，相同时间间隔扫过的扇区数量（字节数）是相同的，这种技术称为恒定角速度（Constant Angular Velocity，CAV）。

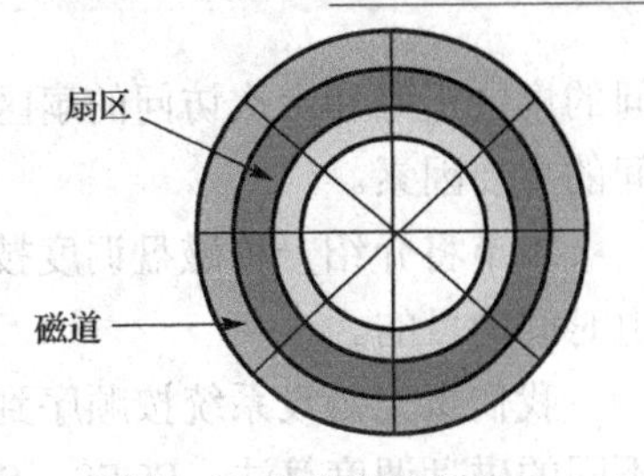

图 8-2 盘片逻辑划分

8.1.2 格式化

格式化（Formatting）是指对磁盘进行的初始化操作，格式化可分为低级格式化（Low-Level Formatting）和高级格式化（High-Level Formatting）。

低级格式化也称物理格式化（Physical Formatting），通常由磁盘的生产厂商完成，在出厂前要将空白盘片初始化成可存储数据状态，也就是对其进行磁道和扇区的划分。为每个扇区定义特殊的数据结构，填充磁盘。每个扇区的数据结构通常由头部、数据区域（通常为 512 字节）及尾部组成。头部和尾部包含了一些磁盘控制器的使用信息，如扇区号和纠错代码（Eror Correcting Code，ECC）。这些工作好比在一个空旷的广场上搭起一间间的板房，每个房间都有编号，大小相等，房间和房间之间有少许间隙防止错位。做完这一步，这些房间如何使用就交给终端用户了。

磁盘存储信息，不但要有物理空间，而且要有一套管理规则，如分区的划分及操作系统放在哪里等。依据这个规则系统才可以准确地存放数据和提取数据。构建规则的过程就是装载文件系统的过程，这就是高级格式化的工作，也称逻辑格式化（Logical Formatting）。操作系统在这步将文件系统数据结构存储到磁盘上，这些数据结构包括空闲、已分配的空间及一个初始为空的目录。

8.2 磁盘调度

8.2.1 磁盘性能指标

根据磁盘的物理特性，磁头读/写数据要经历的过程是：让磁头移动到指定柱面；等待所需扇区旋转到磁头下文；读/写所要的扇区数据。

在整个过程中要耗费三部分时间。

（1）寻道时间 T_s：将磁头定位到正确磁道（柱面）上所花的时间，与盘片直径和传动臂速度相关，平均为 20ms。

（2）旋转延迟 T_r：所查找的扇区转到磁头下所用的时间，与磁盘的旋转速度有关，一个 10000 RPM 磁盘的平均旋转延迟为 3ms。

（3）传送时间 T：传送扇区内数据的时间，同样取决于磁盘的旋转速度，$T = b/(rN)$（b 为要传送的字节数，N 为一个磁道中的字节数，r 为转速）

总的磁盘平均存取时间 $T_a = T_s + T_r + T$。假设磁盘的平均寻道时间为 5ms，平均旋转延迟为 4ms，传输速率是 4MB/s，扇区大小是 1KB，那么如果随机访问一个扇区，总的存取时间 T_a=5+4+0.25=9.25ms；若要访问的扇区和磁头都在同一柱面上，则 T_a=4+0.25=4.25ms；若要访间的扇区正好和上次访问的扇区相邻，则 T_a=0.25ms。可见寻道时间和旋转延迟是影响存取时间的重要因素。

本节将介绍一种磁盘调度技术，可以通过优化磁盘柱面 I/O 请求的响应顺序来达到减少寻道时间的目的。

我们事先假设系统按顺序到达了如下柱面请求：98,183,37,122,14,124,65,67，下面用 5 种不同的磁盘调度算法：FCFS、SSTF、SCAN、C-SCAN 和 LOOK，对其进行调度，计算磁头划过柱面的数量。

8.2.2 FCFS 调度

磁盘调度的最简单算法是先来先服务(First-Come First Served，FCFS)算法，这是一种公平算法，按照柱面 I/O 请求的顺序依次响应。如果磁头开始位于柱面 53，那么它首先从柱面 53 移到柱面 98，接着再到柱面 183、37、122、14、124、65，最后到柱面 67。磁头移动柱面的总数为 640。这种调度如图 8-3 所示。图中的箭头指示的是磁头移动的方向，我们看到了若干次大摆动，说明该调度算法未做任何优化，走了很多回头路，效率低。因此，必须采用一些策略让磁头在移动前做出一个决策，使得移动的距离更短。

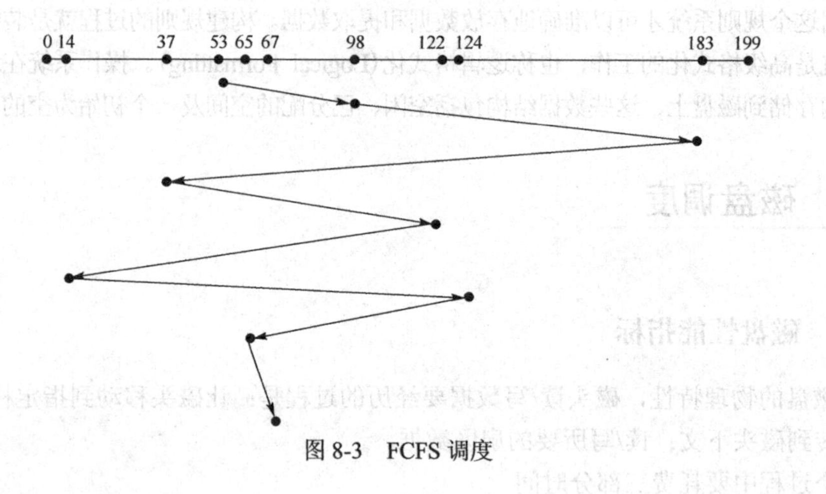

图 8-3 FCFS 调度

8.2.3 SSTF 调度

在移动磁头准备响应下一个请求之前，处理靠近当前磁头位置的所有请求可能较为合理，因为行程最短，这就是最短寻道时间优先(Shortest-Seek-Time-First，SSTF)算法。SSTF 算法选择先处理最接近磁头位置的待处理请求。

接上例，磁头开始位置是柱面 53，离它最近的是柱面 65，所以会先响应它。一旦位于柱面 65，下一个最近的请求位于柱面 67。按同样的策略，后面响应的分别是柱面 37、14、98、122、124、183，移动的柱面总数为 236，约为 FCFS 调度的三分之一，如图 8-4 所示。显然，这种调度大大提高了性能。

SSTF 调度本质上和进程调度的 SJF 算法相同，类似地，SSTF 也会出现饥饿现象。假设磁头正在响应柱面15请求，此时新到达的柱面请求都出现在柱面15附近，按照SSTF的策略会优先响应它们，那么远离柱面15的请求会长时间得不到响应，即饥饿。

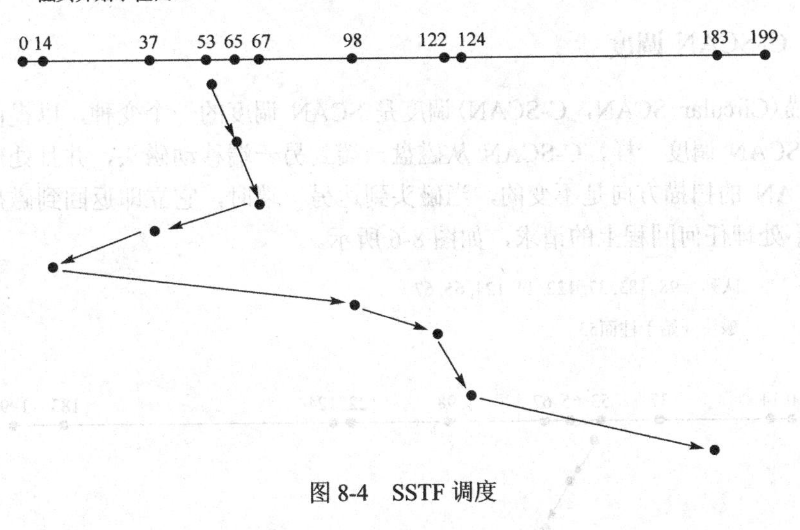

图 8-4　SSTF 调度

8.2.4　SCAN 调度

为了解决饥饿问题，我们需要一种公平算法：扫描算法(SCAN Algorithm)，磁头盘片的一端向另一端移动，在移过每个柱面时响应请求，当到达磁盘的另一端时，磁头方向反转继续处理，就如同来回扫描盘片。

该算法的实践需要知道磁头初始的移动方向，接上例，若磁头此时向柱面 0 方向移动，磁头初始位置还是柱面53，接下来则会处理柱面37，然后处理柱面14。到达柱面0后，磁头方向反转，并处理柱面65、67、98、122、124及183，如图8-5所示。如果一个新的请求柱面刚好在磁头的前进方向前面，那么它很快就会得到响应；如果出现在后面，就必须等到磁头到达端点后反向移动过来。

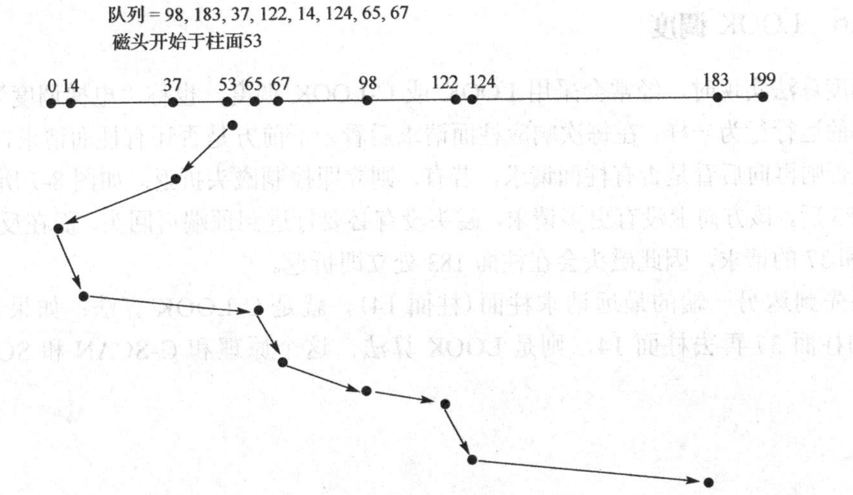

图 8-5　SCAN 调度

假设请求柱面的分布是均匀的，那么磁头刚响应过的柱面请求相对较少，远离磁头的一端请求数量会增加，因此可以考虑磁头到达一端后直接回到起点重新扫描，这就是C-SCAN算法。

8.2.5 C-SCAN 调度

循环扫描(Circular SCAN，C-SCAN)调度是SCAN调度的一个变种，以提供更均匀的等待时间。像SCAN调度一样，C-SCAN从磁盘一端到另一端移动磁头，并且处理行程上的请求，但C-SCAN的扫描方向是不变的，当磁头到达另一端时，它立即返回到磁盘的开头重新扫描，而并不处理任何回程上的请求，如图8-6所示。

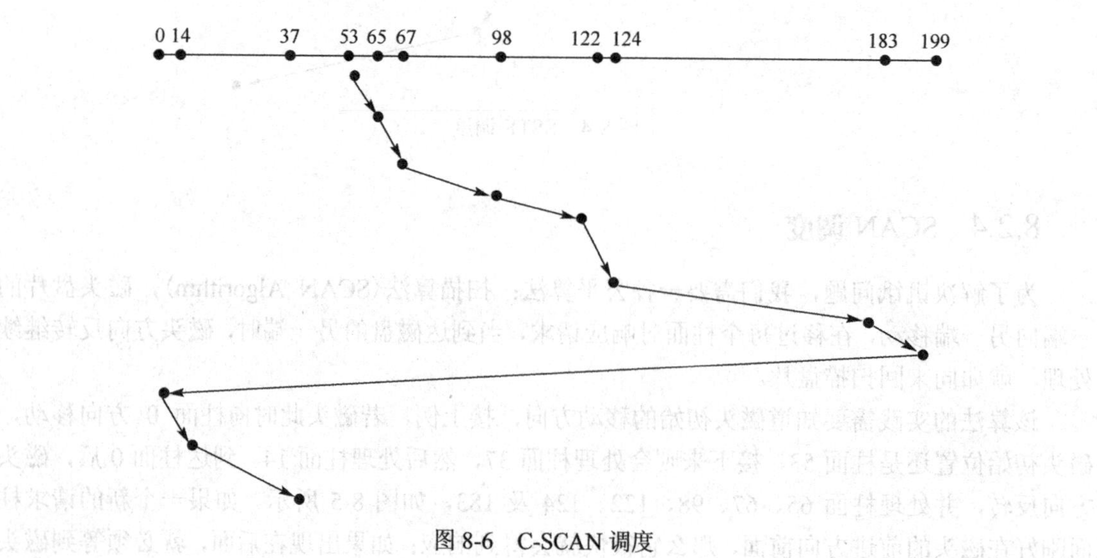

图8-6 C-SCAN调度

8.2.6 LOOK 调度

在调度算法实现时，经常会采用LOOK或C-LOOK调度，也称“电梯调度”。调度过程就像电梯的运行行为一样：在每次响应柱面请求后看一下前方是否还有柱面请求，若有，则继续前进；否则再向后看是否有柱面请求，若有，则立即控制磁头折返。如图8-7所示，在响应完柱面183后，该方向上没有更多请求，磁头没有必要行进到顶端再回头，而在反方向上还有柱面14和37的请求，因此磁头会在柱面183处立即折返。

如果先到达另一端的最远请求柱面(柱面14)，就是C-LOOK算法；如果先到达离磁头最近的柱面37再去柱面14，则是LOOK算法。这个原理和C-SCAN和SCAN是完全类似的。

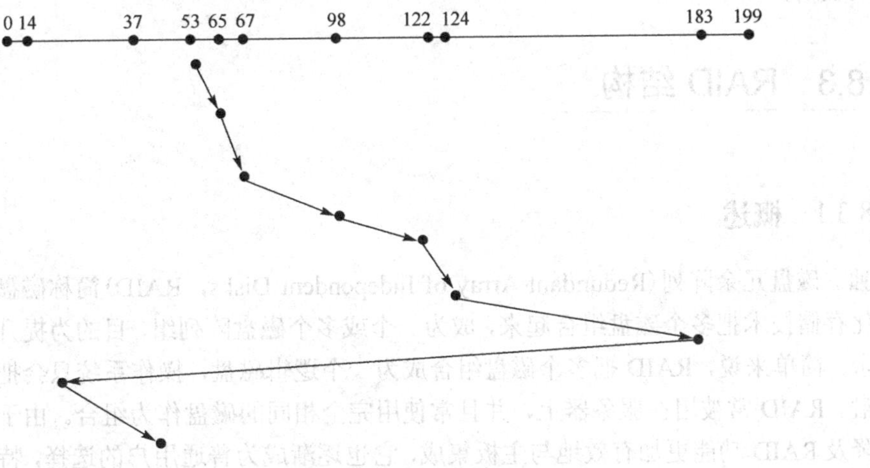

图 8-7　C-LOOK 调度

8.2.7　调度算法选择

给出了如此多的磁盘调度算法，如何选择最佳的算法呢？SSTF 算法是常见的调度算法，并且具有自然的吸引力，因为它比 FCFS 调度具有更好的性能。对于磁盘负荷较大的系统，SCAN 调度和 C-SCAN 调度表现更好，因为它们不太可能造成饥饿问题。对于任何特定的请求列表，可以定义最佳的执行顺序，但是计算最佳调度的所需时间可能得不到补偿。然而，对于任何调度算法，性能在很大程度上取决于请求的数量和类型。例如，假设队列中通常只有一个待处理请求，那么所有调度算法都一样，因为如何移动磁头都只有一个选择。

在 Linux 中，以通过如图 8-8 所示的方法查看系统当前采用的调度策略。图中列出了系统中的三种可用策略：noop、deadline 及 cfq，noop 策略被一对方括号包围，说明当前系统使用的策略是 noop。

```
root@youngyt-PC:/# cat /sys/block/sda/queue/scheduler
[noop] deadline cfq
```

图 8-8　查看 Linux 系统的调度策略

(1) noop：no operation 的简写，它实施最简单的 FCFS 调度算法。不要吃惊，在 SSD（固态驱动器）上执行磁盘调度算法，应当最大限度地减少磁盘驱动器的磁头移动量，通常采用简单的 FCFS 调度算法。

(2) deadline：系统会维护两个队列——读队列和写队列，读队列的请求优先级比写队列的高，为了防止饥饿，每个 I/O 请求都有一个计时器，当计时器时间耗尽时，说明等待时间过长，会将该请求推向最高优先级。

(3) cfq：全称 complete fairness queueing，即完全公平队列。它试图为竞争块设备使用权的所有进程分配一个请求队列和一个时间片，在调度器分配给进程的时间片内，进程可以将其读/写请求发送给底层块设备，当进程的时间片消耗完，进程的请求队列将被挂起，等待调度。每个进程的时间片和每个进程的队列长度取决于进程的 I/O 优先级，每个进程都会有一个 I/O

优先级，cfq 调度器将会将其作为考虑的因素之一，来确定该进程的请求队列何时可以获取块设备的使用权。

8.3 RAID 结构

8.3.1 概述

独立磁盘冗余阵列(Redundant Array of Independent Disks，RAID)简称磁盘阵列。人们用虚拟化存储技术把多个磁盘组合起来，成为一个或多个磁盘阵列组，目的为提升性能或减少资料冗余。简单来说，RAID 把多个磁盘组合成为一个逻辑磁盘，操作系统只会把它当作一个实体磁盘。RAID 常被用在服务器上，并且常使用完全相同的磁盘作为组合。由于磁盘价格的不断下降及 RAID 功能更加有效地与主板集成，它也逐渐成为普通用户的选择，特别是在需要大容量存储空间的工作中，如视频与音频制作。

(1) 可靠性问题的解决方法是引入冗余(Redundancy)，额外信息能够在磁盘故障时用于重建丢失的信息。此时，即使磁盘故障，数据也不会丢失。

(2) 多个磁盘的并行访问改善性能。通过磁盘镜像，读请求的处理速度可以加倍，因为读请求可以被送到任一磁盘中处理。采用多个磁盘，通过将数据分散在多个磁盘上，也可以改善传输速率，这种方法称为并行。例如，如果有 8 个磁盘，则可以将每字节的 8 位分别写到磁盘上。这 8 个磁盘可作为单个磁盘使用，其扇区为正常扇区的 8 倍，更重要的是，它具有 8 倍的访问率，每个磁盘参与每个访问(读/写)，这样每秒所能处理的访问数量与单个磁盘一样，但是每次访问的数据在同样时间内为单个磁盘的 8 倍。

8.3.2 RAID 级别

根据 RAID 级别的不同，资料会以多种模式分散于各个磁盘上，RAID 级别的命名会以 RAID 开头并带数字，如 RAID 0、RAID 1、RAID 5、RAID 6、RAID 7、RAID 01、RAID 10、RAID 50 及 RAID 60。每种级别都有其理论上的优缺点，不同的级别在两个目标间获取平衡，两个目标分别是增加资料可靠性及增强存储器(群)的读/写性能。

1．RAID 0

RAID 0 也称为带区集。它将两个以上的磁盘并联起来，成为一个大容量的磁盘。在存放数据时，将数据分段后分散存储在这些磁盘中。因为读/写时都可以并行处理，所以在所有的级别中，RAID 0 的速度是最快的。但是 RAID 0 既没有冗余功能，也不具备容错能力，如果一个磁盘(物理)损坏，所有数据都会丢失(如图 8-9 所示)。

2．RAID 1

RAID 1 将两组以上的 N 个磁盘相互作为镜像，在一些多线程操作系统中能有很好的读取速度，理论上读取速度等于磁盘数量的倍数，与 RAID 0 相同。另外，其写入速度有微小的降低。其只要一个磁盘正常即可维持运作，可靠性最高。其原理是在主磁盘上存放数据的同时，也在镜像磁盘上存放一样的数据。当主磁盘(物理)损坏时，镜像磁盘则代替主磁盘的工作。因为有镜像磁盘做数据备份，所以 RAID 1 的数据安全性在所有的 RAID 级别中是最高的。但无论用多少磁盘做 RAID 1，仅算一个磁盘的容量，所以其是所有 RAID 级别中磁盘利用率最低的。如果用两个不同大小的磁盘做 RAID 1，可用空间为较小的那个磁盘，较大的磁盘多出来的空间也可以分割成一个区来使用，不会造成浪费(如图 8-10 所示)。

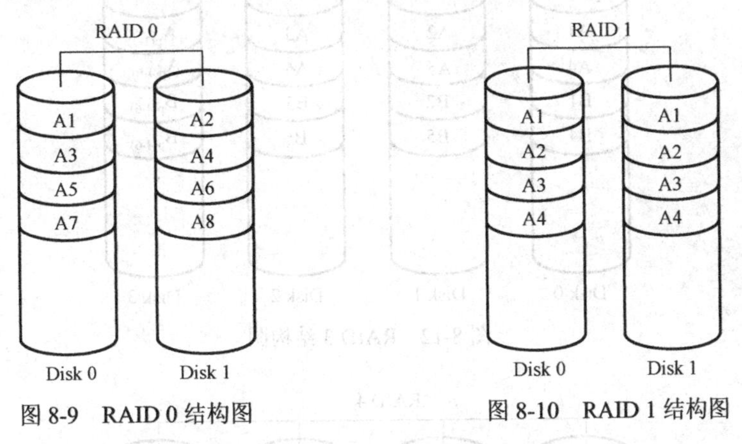

图 8-9　RAID 0 结构图　　图 8-10　RAID 1 结构图

3．RAID 2

RAID 2 这是 RAID 0 的改良版，以汉明码(Hamming Code)的方式将数据进行编码后分割为独立的位，并将数据分别写入磁盘中。因为在数据中加入了错误修正码(Error Correction Code，ECC)，所以数据整体的容量会比原始数据大一些(如图 8-11 所示)。

RAID 2 最少需要三台磁盘驱动器方能运作。

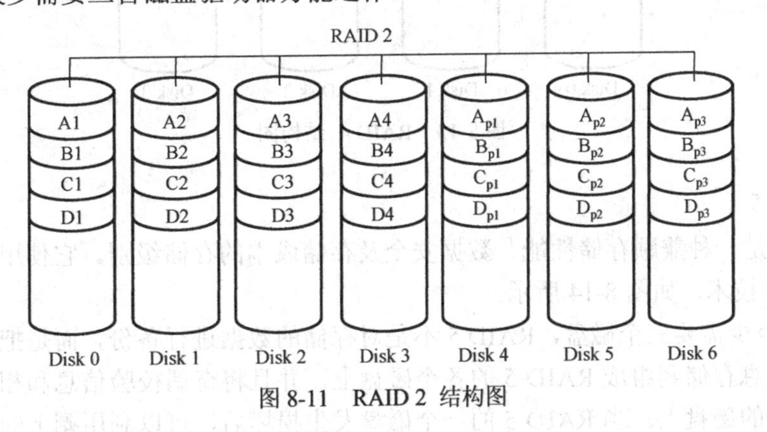

图 8-11　RAID 2 结构图

4．RAID 3

RAID 3 采用数据交织存储(Bit-interleaving)技术，它需要先通过编码，将数据位分割后分别存在磁盘中，而将同位检查后的数据单独存在一个磁盘中，但由于数据内的位分散在不同的磁盘上，因此就算要读取一小段数据，都可能需要所有的磁盘进行工作，所以这种级别比较适用于读取大量数据(如图 8-12 所示)。

5．RAID 4

如图 8-13 所示，RAID 4 与 RAID 3 不同的是，它在分割数据时是以区块为单位的，然后分别将它们存在磁盘中(块交织存储，Block-interleaving)，但每次的数据访问都必须从同位检查的那个磁盘中取出对应的同位数据进行核对，因为过于频繁地使用磁盘，所以对磁盘的损耗可能会提高。

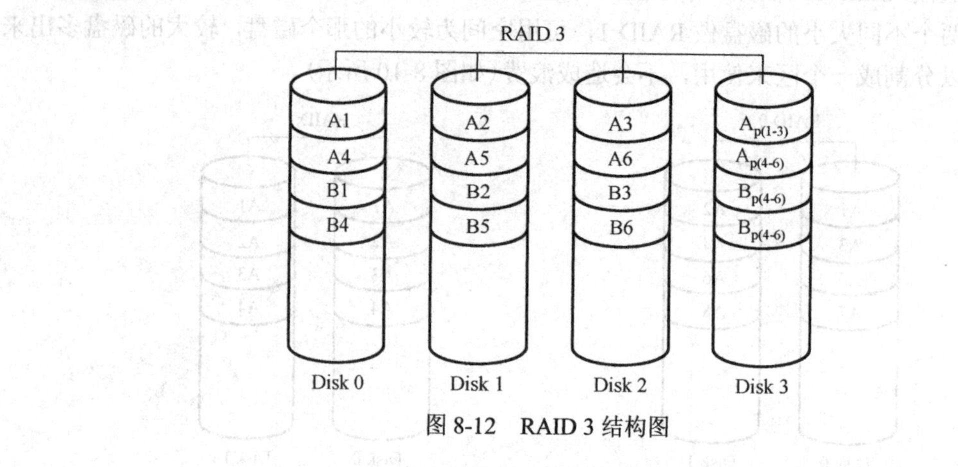

图 8-12　RAID 3 结构图

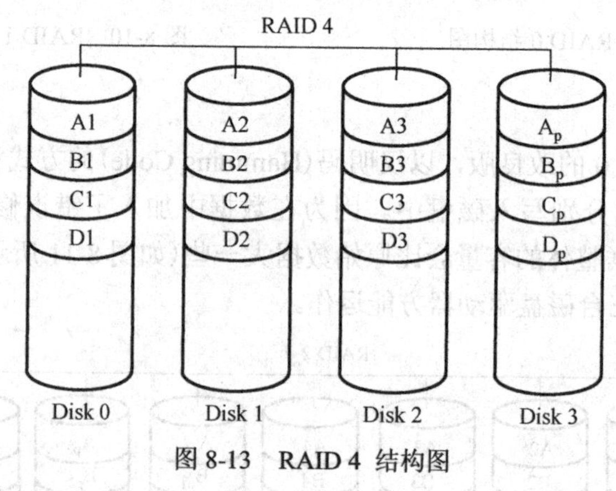

图 8-13　RAID 4 结构图

6. RAID 5

RAID 5 是一种兼顾存储性能、数据安全及存储成本的存储级别。它使用的是磁盘分割(Disk Striping)技术，如图 8-14 所示。

RAID 5 至少需要三个磁盘，RAID 5 不是对存储的数据进行备份，而是把数据和相对应的奇偶校验信息存储到组成 RAID 5 的各个磁盘上，并且将奇偶校验信息和相对应的数据分别存储于不同的磁盘上。当 RAID 5 的一个磁盘发生损坏后，可以利用剩下的数据和相应的奇偶校验信息去恢复被损坏的数据。RAID 5 可以理解为 RAID 0 和 RAID 1 的折中方案。RAID 5 可以为系统提供数据安全保障，但保障程度要比镜像低，而磁盘空间利用率要比镜像高。RAID 5 具有和 RAID 0 相似的数据读取速度，只因为多了一个奇偶校验信息，写入数据的速度相对单独写入一个磁盘的速度略慢，若使用“回写缓存”，则可以让性能改善不少。同时由于多个数据对应一个奇偶校验信息，RAID 5 的磁盘空间利用率要比 RAID 1 高，存储成本相对较低。

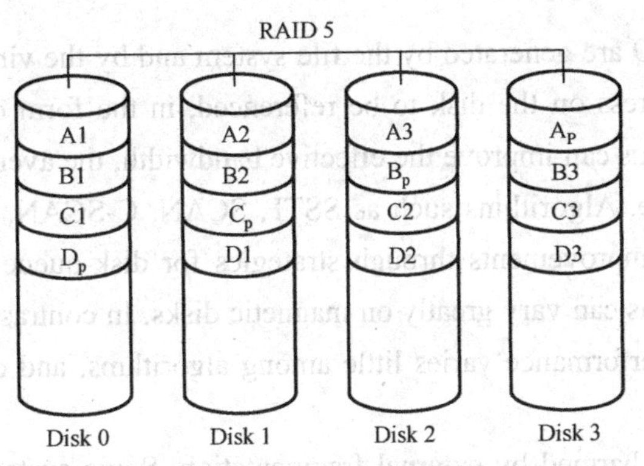

图 8-14 RAID 5 结构图

7. RAID 6

与 RAID 5 相比，RAID 6 增加第二个独立的奇偶校验信息。两个独立的奇偶校验系统使用不同的算法，数据的可靠性非常高，任意两个磁盘同时失效时，都不会影响数据完整性。RAID 6 需要分配给奇偶校验信息更大的磁盘空间和额外的校验计算时间，相对于 RAID 5 有更大的 I/O 操作量和计算量，其写性能取决于具体的实现方案，因此 RAID 6 通常不会通过软件方式来实现，而更可能通过硬件方式实现(如图 8-15 所示)。

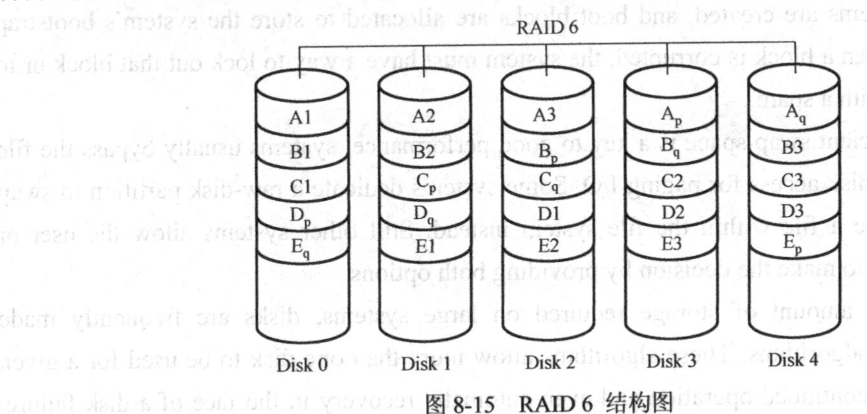

图 8-15 RAID 6 结构图

同一数组中最多容许两个磁盘损坏。更换新磁盘后，资料将会重新被写入新的磁盘中。RAID 6 在硬件磁盘阵列卡的功能中，也是最常见的磁盘阵列等级。

8.4 Reading Materials

Disk drives are the major secondary storage I/O devices on most computers. Most secondary storage devices are either magnetic disks or magnetic tapes, although solid-state disks are growing in importance. Modern disk drives are structured as large one-dimensional arrays of logical disk blocks. Generally, these logical blocks are 512 bytes in size. Disks may be attached to a computer system in one of two ways: (1) through the local I/O ports on the host computer or (2) through a network connection.

Requests for disk I/O are generated by the file system and by the virtual memory system. Each request specifies the address on the disk to be referenced, in the form of a logical block number. Disk-scheduling algorithms can improve the effective bandwidth, the average response time, and the variance in response time. Algorithms such as SSTF, SCAN, C-SCAN, LOOK, and C-LOOK are designed to make such improvements through strategies for disk-queue ordering. Performance of disk-scheduling algorithms can vary greatly on magnetic disks. In contrast, because solid-state disks have no moving parts, performance varies little among algorithms, and quite often a simple FCFS strategy is used.

Performance can be harmed by external fragmentation. Some systems have utilities that scan the file system to identify fragmented files; they then move blocks around to decrease the fragmentation. Defragmenting a badly fragmented file system can significantly improve performance, but the system may have reduced performance while the defragmentation is in progress. Sophisticated file systems, such as the UNIX Fast File System, incorporate many strategies to control fragmentation during space allocation so that disk reorganization is not needed.

The operating system manages the disk blocks. First, a disk must be lowlevel-formatted to create the sectors on the raw hardware, new disks usually come preformatted. Then, the disk is partitioned, file systems are created, and boot blocks are allocated to store the system's bootstrap program. Finally, when a block is corrupted, the system must have a way to lock out that block or to replace it logically with a spare.

Because an efficient swap space is a key to good performance, systems usually bypass the file system and use raw-disk access for paging I/O. Some systems dedicate a raw-disk partition to swap space, and others use a file within the file system instead. Still other systems allow the user or system administrator to make the decision by providing both options.

Because of the amount of storage required on large systems, disks are frequently made redundant via RAID algorithms. These algorithms allow more than one disk to be used for a given operation and allow continued operation and even automatic recovery in the face of a disk failure. RAID algorithms are organized into different levels; each level provides some combination of reliability and high transfer rates.

第9章 文件管理

获取视频

9.1 概述

在上一章中我们讨论了磁盘的构造及两种格式化方式，其中低级格式化是为了给磁盘划分出可供存储信息的单元(扇区或物理块)，而高级格式化是为了装载文件系统。我们把磁盘想象成一个巨大的仓库，低级格式化将仓库分隔成了一间间相同大小的房间，每个房间都有一个编号(从0开始编号)。货物就存放在这些房间内，如果一个房间装不下，就将货物拆分，放在不同的房间中。货物的最后一部分可能没有办法占满整个房间，那么这间房的剩余空间就是碎片。拆分的货物如果放在连续编号的房间中，那只要知道占用的第一个房间编号和货物大小即可，如果分散存放，那就要用额外的信息来记录它们的位置。

回到现实中来，因为磁盘很大，为了更有效地管理空间，可能需要对它先进行分区，在分区中以目录的形式管理数据，数据在磁盘上的呈现形式就是文件。为了完成上述管理工作，设计人员开发了一套管理软件，在安装操作系统时一起装在磁盘当中，这套软件就是本章的“主角”——文件系统(File System)，本章主要的话题有两个：文件(File)和目录(Directory)。

对大多数用户来说，和操作系统交互最多的子系统就是文件系统。他们每天可能会创建、编辑、保存或删除文件，这些对文件的操作都是由文件系统完成的，文件系统负责存取和管理信息资源，采用统一的方法管理用户信息和系统信息的存储、检索、更新、共享及保护，并为用户提供一整套有效的文件使用及操作方法。

信息是以文件为单位存储在外存上的，这一点与内存管理不同，内存管理的目标硬件是内存，存储单位是字节，存取靠的是内存地址。

9.2 文件

文件是信息的逻辑存储单位。在用户看来，文件中的内容是具有逻辑结构的信息集合，如一篇文章、一段代码，都是逻辑上可读的。但对于系统而言，文件的本质是存储在外存中的二进制数据集合。因此，文件系统既要满足用户的逻辑需求，也要考虑到文件存储的物理需求。

文件由操作系统映射到存储设备上。这些存储设备通常是非易失性的，因此在系统重新启动之后，文件内容依旧存在。

文件信息由创建者定义。文件可存储许多不同类型的信息，如源代码或可运行程序，数字或文本数据，照片、音乐及视频等。文件具有某种结构，这取决于其类型。文本文件(Text File)为按行(可能还有页)组织的字符序列；源文件为函数序列，而每个函数包括声明和可执行语句；可执行文件(Executable File)为一系列代码段，以供加载程序调入内存并运行。

9.2.1 文件类型

文件类型用于指示文件内部结构，操作系统通过文件类型决定对文件进行何种解释。比如，对可执行文件，操作系统会尝试加载执行；对文本文件，操作系统会将其内容展示出来等。

为了区别文件类型，将文件类型作为文件名的一部分。文件名分为两部分，即文件名称和扩展名称，通常由句点分开，如“备注.txt”，这样，用户和操作系统仅从文件名就能得知该文件是文本类型文件。常用的文件类型如表9-1所示。

表9-1 常用的文件类型

文件类型	常用扩展名称	功 能
可执行文件	.exe，.com	可运行的机器语言程序
目标文件	.obj，.o	已编译的、尚未链接的机器语言代码
源代码文件	.c，.cc，.java，.py	各种语言的源代码
批处理文件	.bat，.sh	解释程序的指令
标记语言文件	.xml，.html	文本数据、文档
归档文件	.rar，.zip，.tar	相关文件组成的一个文件，用于归档或存储
图像文件	.gif，.pdf，.jpg	打印或图像格式的ASCⅡ码或二进制文件

对于Linux系统，文件类型是通过在文件最开始的部分写入一些特征值来确认的，这些特征值称为幻数(Magic Number)，如可执行文件、shell脚本及PDF文件等，当然并不是所有文件都有幻数。而文件名中的扩展名称仅作为对用户的一种提示，Linux既不强制、也不依赖这些扩展名称。因为文件类型繁多，操作系统不可能对所有类型都加以解释，所以Linux系统本身只能解释两种文件类型：文本文件及二进制可执行文件。它认为每个文件由字节序列构成，解释这些字节的工作由对应的应用程序完成。比如，对.doc文件，操作系统本身是无法识别的，但是如果系统中安装了WPS Office应用程序，那么会将该文件交给WPS Office打开解释，文件类型和应用程序之间会建立一种打开关联。所以在面对未知文件类型且无相关联应用时，系统一般会提示让用户选择应用程序打开，或者默认按文件方式打开。

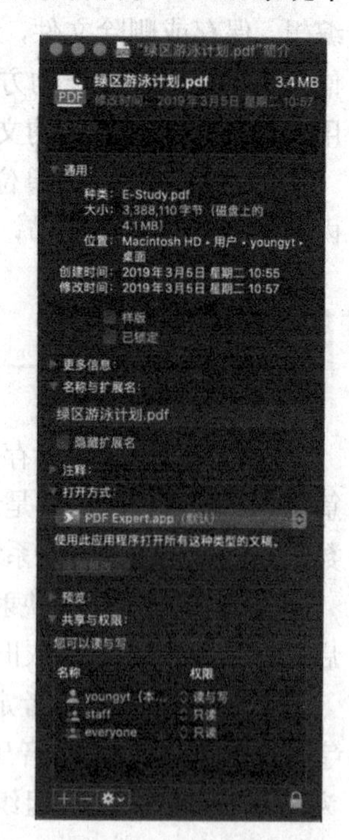

图9-1 文件属性

9.2.2 文件属性

文件是“按名存取”的，因此文件名是文件最重要的属性。有的操作系统区分文件名中的大小写字母，如UNIX、Linux及macOS；而有的操作系统则不区分，如Windows。文件当然还包括其他属性，因操作系统而异，如图9-1所示的是macOS下的文件属性。

(1)文件名：文件名是以可读形式来保存的唯一信息。

(3)种类：即文件类型，支持不同类型文件的系统需要这个属性。

(4)位置：该属性为指向设备与设备上文件位置的指针。

(5)大小：该属性包括文件的当前大小(以字节、字或块为单位)及可能允许的最大大小。

(6)共享与权限：该属性为访问控制信息，确定能进行读取、写入等操作的对象。

(7)时间、日期和用户标识：该属性为文件创建、最后修改和最后使用的相关信息。这些数据用于保护和监控文件。

9.3 存取方法

存取方法是指读/写文件物理记录的方法，根据文件类型的不同，用户使用的要求也不同，操作系统要提供多种存取方法来满足用户要求。

9.3.1 顺序存取

最简单的文件存取方法是顺序存取(Sequential Access)，即按文件信息的顺序处理，文件可以是无结构的字节流，也可以是有结构的记录文件(如数据库文件是由一条条记录构成的)。在读此类文件时，读取完当前文件指针所指向的数据内容后，文件指针自动向后移动；在写文件时，在当前文件指针位置增加内容，并将文件指针移动到文件的结尾。

磁带模型是顺序存取的典型代表，有的系统支持向前或向后跳过若干字节(或记录)进行读/写。

9.3.2 直接存取

直接存取(Direct Access)也称随机存取(Random Access)，该方法可以直接定位到文件中的所需内容处并进行读/写操作。直接存取方法基于文件的磁盘模型，磁盘的存储单位是扇区，扇区按顺序编号且大小相等，如果存放文件的第一个扇区号是0，那么第N个扇区的位置可以直接计算出来，是从起始位置开始第$L\times N$字节的地方(L是每个扇区的大小)。直接存取方法对读/写的顺序没有限制。对于大量信息的立即访问，直接存取方法极为有用。例如，当数据库需要查询特定主题时，首先计算哪个扇区包含结果，然后直接读取相应扇区以提供期望的信息。

并非所有操作系统都支持文件的顺序存取和直接存取。有的操作系统只允许顺序存取，有的操作系统只允许直接存取。因为有的操作系统要求在创建文件时将其定义为顺序的或直接的，所以这样的文件只能按与声明一致的方式来访问。

9.4 目录

获取视频

9.4.1 基本概念

如图9-2所示，为了有效管理空间，对磁盘可以进行分区，每个分区的开始部分放置目录，目录就是一张表格，里面的每个条目记录着文件名和与之对应的属性及磁盘物理位置。操作系统也支持将几个磁盘的空间合并成一个独立的分区(卷)来使用，目录同样位于分区的最开始部分。目录可理解为符号表，对目录的操作包括插入条目、删除条目、搜索给定条目及列出所有或指定条件的条目等，对应的目录功能如下。

(1) 查找文件：通过给定的文件名在目录中查找文件是否存在。

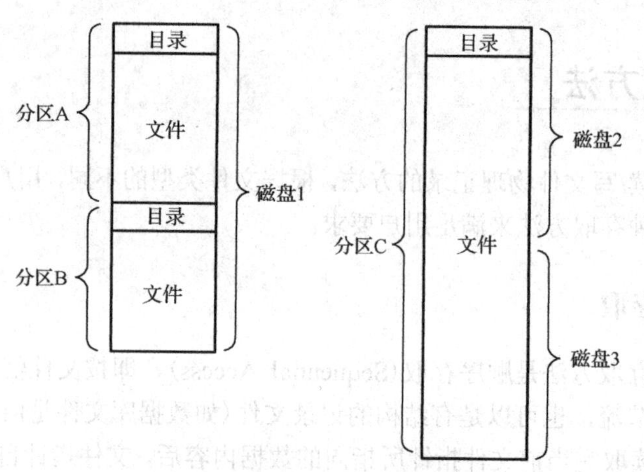

图 9-2　典型的文件系统结构

(2) 创建文件：在磁盘中创建新文件，并将其增加到目录中。

(3) 删除文件：删除目录中对应的条目。

(4) 列出目录：列出目录中所有文件及其对应属性。

(5) 重命名：改变文件名。

(6) 遍历文件系统：遍历文件系统中所有的目录及目录下的文件。

9.4.2　文件控制块

一个典型的目录结构如图 9-3 所示，目录中有若干条目，每个条目对应一个文件，每个条目分成两部分：第一部分是文件名，第二部分是用于描述和控制文件的数据结构，称为“文件控制块”(File Control Block，FCB)。为方便讨论，我们将两部分合并称为 FCB，那么目录就是由一系列 FCB 构成的。目录虽然不保存文件的具体内容，但也要占用磁盘空间，有趣的是，目录也是以文件的形式存储的，存放目录的文件又称为“目录文件”。文件控制块通常含有三类信息：基本信息、存取控制信息及使用信息。

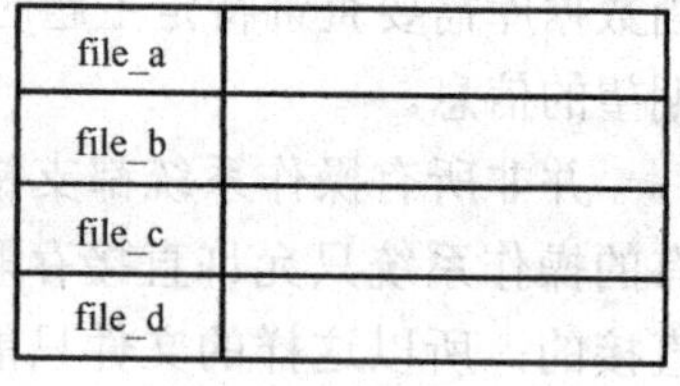

图 9-3　目录结构

1. 基本信息

基本信息包括如下信息。

(1) 文件名，指用于标识一个文件的符号名。在每个系统中，每个文件都必须有唯一的名称，用户利用该名称对其进行存取。

(2) 文件物理位置，指文件在外存上的存储位置，包括存放文件的设备名、文件在外存上的起始盘块号及文件所占用的盘块数或字节数。

(3) 文件逻辑结构，标识文件是流式文件还是记录式文件。

(4) 记录数：标识文件是定长记录式文件还是变长记录式文件等。

(5) 文件的物理结构，标识文件是顺序文件还是链接文件或索引文件。

2. 存取控制信息

存取控制信息包括文件的存取权限、核准用户的存取权限及一般用户的存取权限。

3. 使用信息

使用信息包括文件的创建日期和时间、文件上一次修改的日期和时间及当前使用信息(这项信息包括当前已打开文件的进程数、文件是否被其他进程锁住及文件在内存中是否已被修改但尚未复制到磁盘上)。应该说明，不同操作系统的文件系统，由于功能不同，可能只含有上述信息中的某些部分。

9.4.3 单级目录

如图 9-4 所示，单级目录是最简单的目录结构，是指在整个文件系统中只建立一张目录表，每个文件占一个目录项，目录项中含文件名、文件长度、文件类型、文件物理地址及其他文件属性。

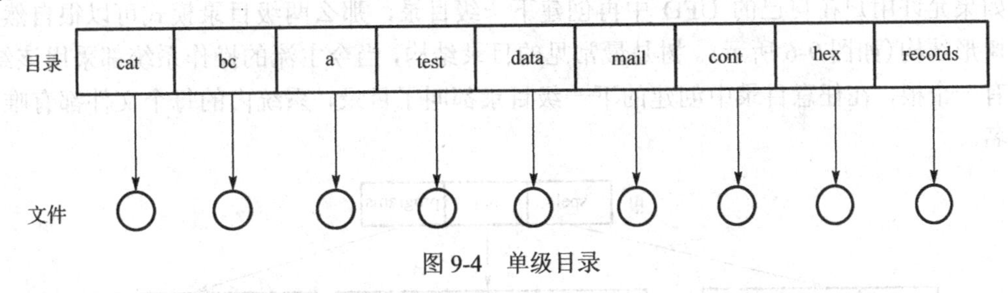

图 9-4 单级目录

然而，当文件数量增加或系统中有多个用户时，单级目录的使用会受到限制。因为所有文件位于同一目录中，它们必须具有唯一的名称。随着文件数量的增加，即使单级目录的单个用户也会难以记住所有文件名称，跟踪这么多的文件是个艰巨的任务。

9.4.4 两级目录

使用单级目录常常会产生混乱的文件名，常用的解决方案是为每个用户创建单独的目录。对于两级目录结构，每个用户都有自己的文件目录(User File Directory，UFD)。这些 UFD 具有类似的结构，但是只列出了单个用户的文件。主文件目录(Master File Directory，MFD)保存用户与其 UFD 的对应关系，当用户登录时会搜索系统的 MFD，通过用户名索引 MFD 找到该用户的 UFD，如图 9-5 所示。

用户操作文件都是在自己 UFD 范围内进行的，相当于和其他用户的文件是隔离开的，可解决名称冲突的问题。但是这种隔离给用户间的合作带来了困扰，他们无法与其他用户共享自己的文件。

为了实现互访，出现了文件路径(File Path)。图 9-5 中的 test 文件，它的路径就是“/user1/test”。为了让 user2 可以访问到该文件，就要修改文件系统的操作逻辑：对于指定路径的文件，先以路径的第一部分为索引搜索 MFD，找到对应的 UFD，再以路径第二部分搜索该 UFD 下的文件。用于搜索给定名称的文件所用的目录序列称为搜索路径(Search Path)。对于给定的指令名称，搜索路径可以扩展到包含需要搜索的无限的目录列表中。这种方法是 UNIX 和 Windows 最常用的。操作系统也可以设计成让每个用户都有自己的搜索路径。

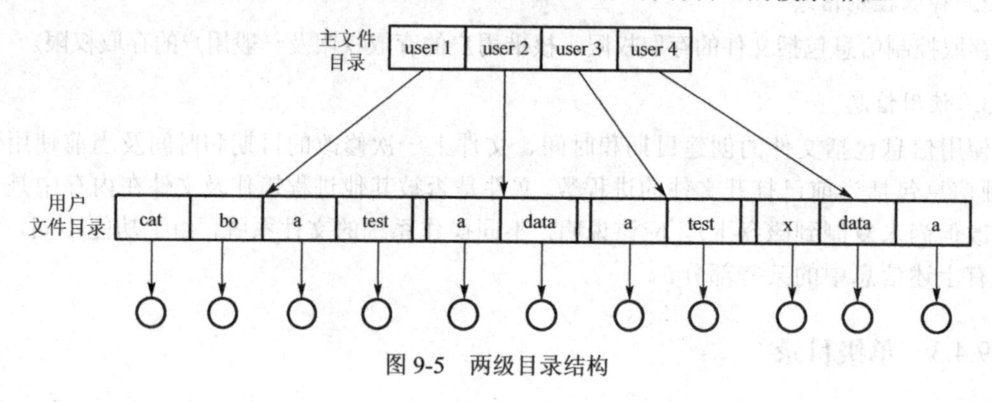

图 9-5　两级目录结构

9.4.5　树形目录

如果允许用户在自己的 UFD 中再创建下一级目录，那么两级目录模式可以很自然地扩展成树形结构(如图 9-6 所示)。树是最常见的目录结构，当今主流的操作系统都采用该结构。树中有一个根，在任意目录中创建的下一级目录都叫子目录，系统内的每个文件都有唯一的路径名。

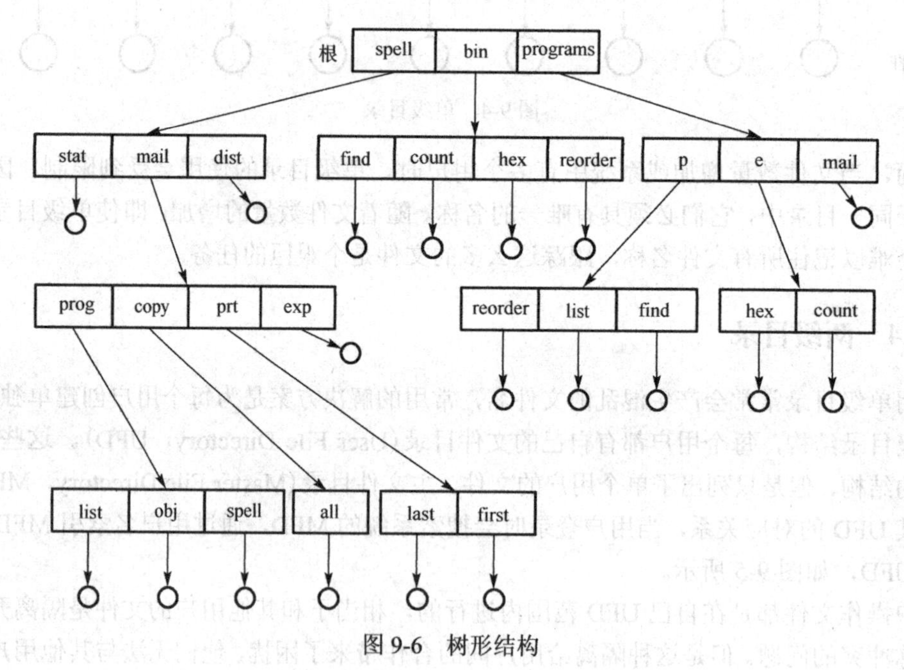

图 9-6　树形结构

路径名有两种形式：绝对路径名和相对路径名。绝对路径名(Absolute Path Name)从根开始，遵循一个路径到指定文件，并给出路径上的目录名。相对路径名(Relative Path Name)从当前目录开始定义一个路径。例如，在图 9-6 所示的树形结构中，如果当前目录是 root/spell/mail，则相对路径名为 prt/first，绝对路径名为 root/spell/mail/prt/first。

采用树形结构，用户除可以访问自己的文件外，还可以访问其他用户的文件。例如，用户B可以通过指定用户A的路径名，来访问用户A的文件，或者用户B可以将其当前目录改为用户A的目录，进而直接采用文件名来访问文件。

9.4.6 UFS的目录实现

UNIX File System(UFS)是UNIX系统中的文件系统格式，在UFS中，每个目录项包括两部分：14字节的文件名和2字节的inode编号。inode被称为索引节点，就是之前我们说的FCB，它包含文件的元数据(Metadata)，如图9-7所示。为了不让每个目录项过于臃肿，UFS在磁盘中开辟了一块专门区域来存放inode，目录项中只记录inode的编号，在图9-8所示的例子中，访问文件“C.file”的过程如下：通过文件名在目录中找到对应项并得到其inode编号12；在磁盘inode区中找到12号inode数据inode-12；从inode-12中读取文件的磁盘位置(61号扇区)；最后向系统发出读/写61号扇区的I/O请求。

inode
- permission
- date&time
- owner/group/ACL
- file size
- ...
- an array of 15 disk-block address

图9-7 inode内容

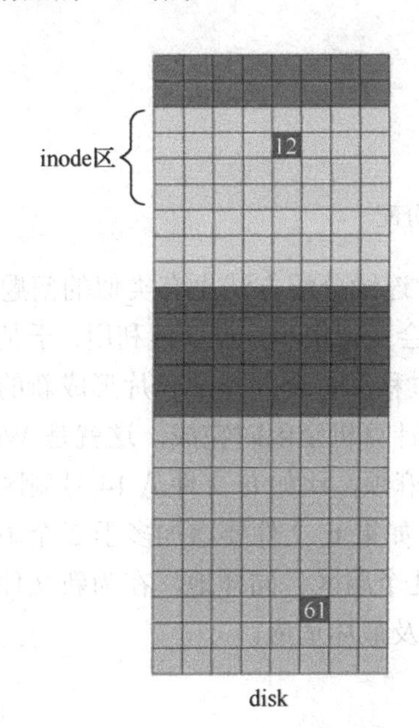

directory

A. file	13
B. file	11
C. file	12

inode-12
- permission
- date & time
- owner/group/ACL
- file size
- ...
- 61(disk-block address)

图9-8 UFS文件存取过程

9.5 分配方法

分配方法可理解为文件的物理结构，即文件内容在磁盘上的物理结构是怎样的。本节共讨论三种为文件分配磁盘空间的方法：连续分配、链接分配及索引分配，我们假定磁盘的可用扇区号从0开始线性递增。

9.5.1　连续分配

连续分配(Contiguous Allocation)方法要求每个文件在磁盘上占用一组连续的扇区。如图 9-9 所示，目录中有 4 个条目，每个条目包含文件名、起始扇区号及文件长度(单位：扇区数)。在存取文件时，只要读出该文件的起始扇区号和长度就可以访问文件的全部内容。这种方法使得文件内容连续，可以避免磁头在不同柱面上移动，使寻道时间最短。

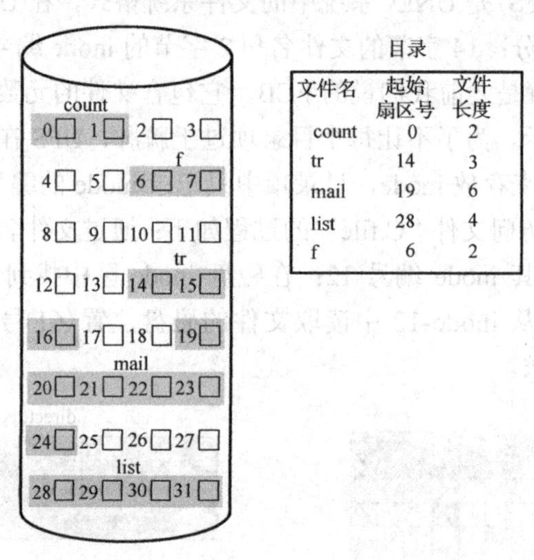

图 9-9　连续分配

回想一下学过的内存分配中提到的外部碎片，连续分配方法也有类似的问题。随着时间的推移，磁盘中文件和文件之间的可用扇区数可能会变得很少，难以再利用，于是就形成了外部碎片，解决外部碎片的方法仍然是“紧缩”，通过移动、合并外部碎片形成新的可用空间，但是这种操作比内存“紧缩”更费时间，通常会在计算机空闲时完成，这就是 Windows 的磁盘碎片整理功能。除此之外，新增文件内容也会有麻烦，比如 tr 文件从 14 号扇区开始占用了 3 个扇区，它和 mail 文件之间只有 2 个空闲扇区，如果 tr 文件要增加多于 2 个扇区的数据内容，就得另想办法，比如将整个 tr 文件向前移动几个扇区。同样地，在为新文件进行空间分配时，可选择不同的策略，如首次适应、最好适应及最坏适应。

9.5.2　链接分配

我们同样可以用离散存储的思想来解决连续分配遇到的问题，其中一个解决方法是链接分配(Linked Allocation)，如图 9-10 所示。文件占用的扇区可以分散在磁盘的任何地方，文件扇区之间用指针链接在一起，一种可行的方案是在每个文件扇区的最后记录下一个扇区的编号，最后一个文件扇区记录一个结束标记。相应地，在目录条目中包含文件名、起始扇区号及结束扇区号，图中 jeep 文件起始于 9 号扇区，9 号扇区中记录的下一个扇区是 1 号扇区，……，最后是 25 号扇区(结束标记是–1)。

采用链接分配方式没有外部碎片，空闲空间列表中的任何块可以用于满足请求。当创建文件时，并不需要说明文件长度，只要有可用的空闲扇区，文件就可以继续写入，因此无须合并磁盘空间。

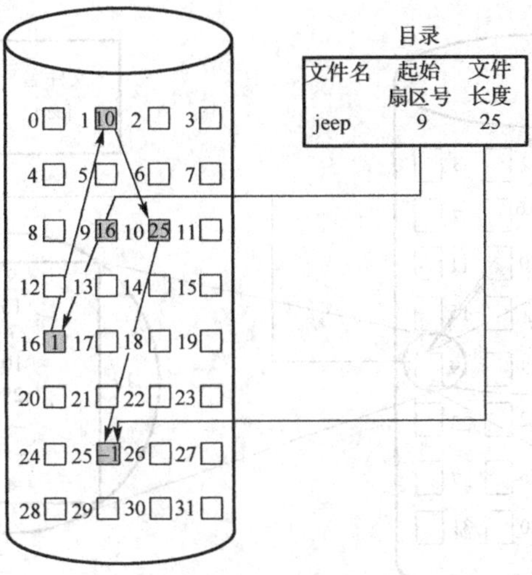

图 9-10　链接分配

根据链表的特点，采用链接分配的文件只能被顺序存取，用户访问文件中的任意内容都要从文件的起始扇区号开始进行遍历，无法直接访问。链表结构的可靠性也是个问题，若其中一个指针损坏，则文件的内容都会丢失，无法找回。

文件分配表(File-Allocation Table，FAT)是链接分配的重要变种，从 MS-DOS 到 Windows 系统都在使用。文件扇区之间的链接关系不存储在扇区最后，而另外用一张表格记录，这张表就叫 FAT 表。由 217、618 及 339 号扇区组成的文件的 FAT 表，如图 9-11 所示。

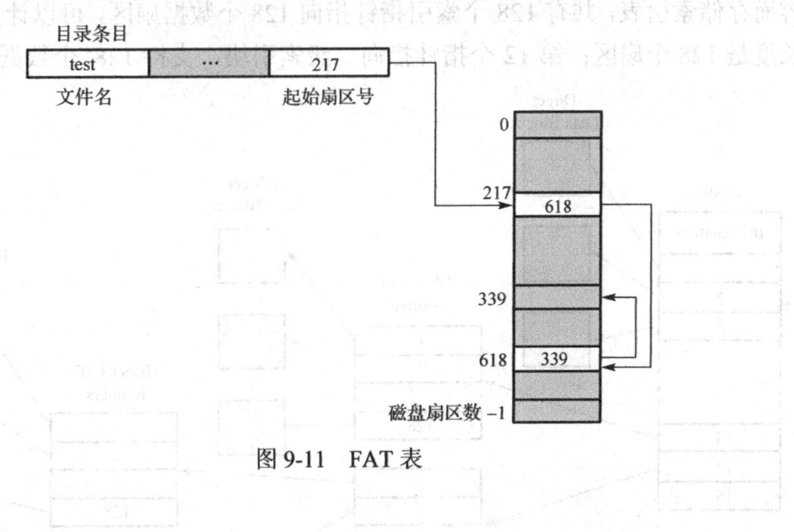

图 9-11　FAT 表

9.5.3　索引分配

链接分配虽然解决了连续分配的外部碎片问题，然而它只支持顺序存取文件。索引分配(Indexed Allocation)通过将所有指针放在索引块(Index Block)中实现高效的直接存取。

如图 9-12 所示，目录中只记录文件名和索引块位置，存取文件时先读出索引表，索引表中记录了文件所有扇区的编号和顺序，无须像链表那样遍历，就可以直接访问文件的任意部分。索引分配支持直接存取，并且没有外部碎片，索引分配中的索引表是额外的存储开销，而且通常大于链接分配中的指针开销。

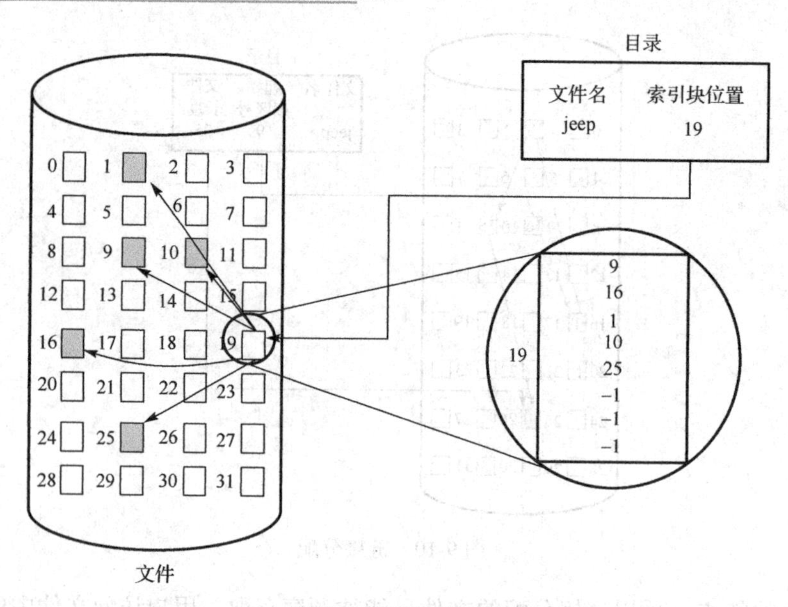

图 9-12　索引分配

UFS 采用了一种三级索引分配方案，有效地解决了小文件额外开销大的问题，如图 9-13 所示。假设扇区大小为 512 字节，每个索引指针占 4 字节，UFS 的 inode 中包含 13 个索引指针：前 10 个指针指向数据扇区，10 个扇区大小以内的文件可以直接索引；第 11 个指针指向一级索引块，如果文件长度超过 10 个扇区，那么就用一级索引，这个指针指向的扇区不存储文件内容而存储索引表，共有 128 个索引指针指向 128 个数据扇区，可以计算出一级索引支持的文件长度是 128 个扇区；第 12 个指针指向二级索引块，支持 128^2 个数据扇区；第 13 个指

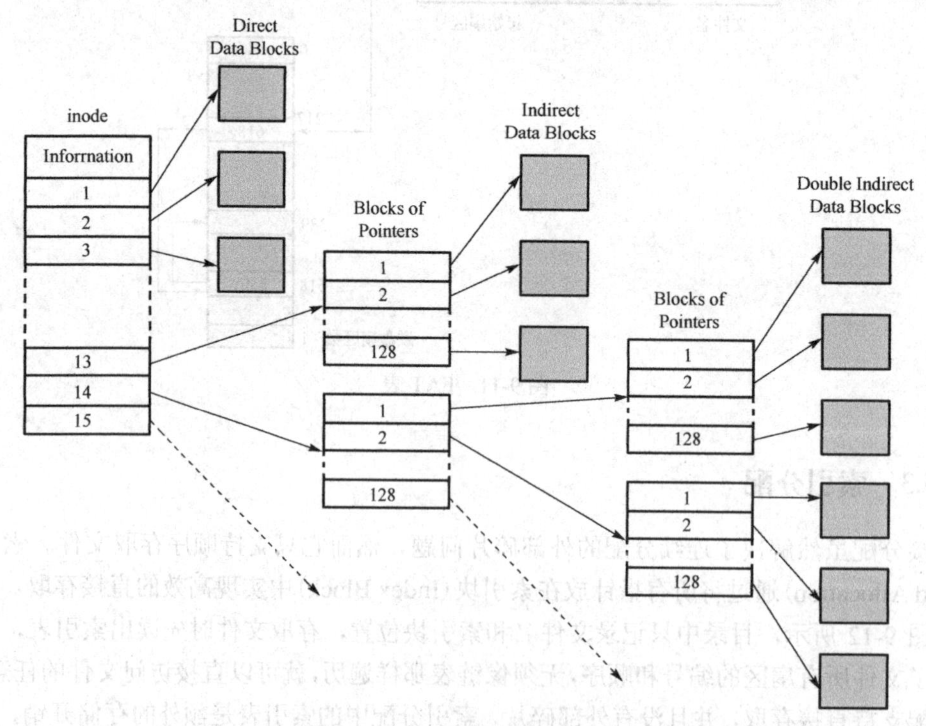

图 9-13　UFS 三级索引分配方案

针指向三级索引块，支持 128^3 个数据扇区。该方案对于直接查找小文件的方式，可减少索引块的开销；对于大文件，则以多级索引的方式来支持，所以大文件在访问扇区时需要进行大量查询。该方案也应用在了 Linux 的 ex2/ex3 文件系统里，虽然解决了大文件存储问题，但是查询效率依旧很低，在最新的 ex4 文件系统中，这个问题得到了改进。

9.6 空闲空间管理

9.6.1 位图法

通常，空闲空间列表由位图（Bit Map）或位向量（Bit Vector）来实现。每个扇区的状态用一位来表示：0 表示已分配，1 表示空闲，如图 9-14 所示。这种方法的主要优点是，在查找磁盘上的第一个空闲扇区和 n 个连续的空闲扇区时相对简单高效。但因为每次分配都要查询位图，所以如果位图在内存中，检索效率会比较高，但是大容量磁盘位图也是比较大的，很难全部放入内存（如具有 4 KB 扇区的 1 TB 磁盘需要 256 MB 的空间来存储位图）。

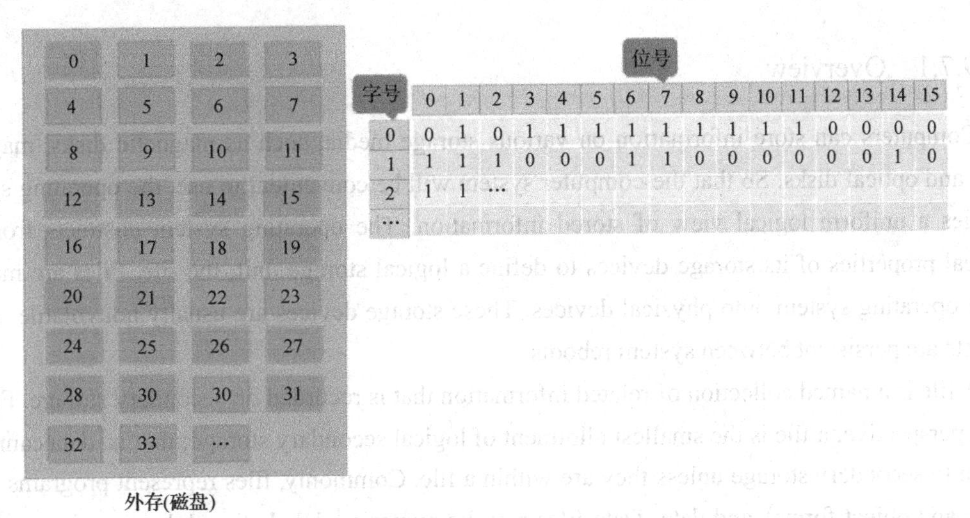

图 9-14　位图法

9.6.2 空闲链表法

空闲空间管理的又一种方法是空闲链表法，如图 9-15 所示。将所有空闲盘块用链表链接起来，将指向第一个空闲盘块的指针保存在磁盘的特殊位置上，同时也将其缓存在内存中。分配时，从链头开始依次取出空闲盘块分配，并修改空闲链表的链头指针；回收时，将回收的盘块依次写入链尾，并修改空闲链表的链尾指针。幸运的是，遍历空闲链表不是一个频繁操作，因为其通常只使用空闲链表的第一块。

图 9-15　空闲链表法

9.7 Reading Materials

9.7.1 Overview

Computers can store information on various storage media, such as magnetic disks, magnetic tapes, and optical disks. So that the computer system will be convenient to use, the operating system provides a uniform logical view of stored information. The operating system abstracts from the physical properties of its storage devices to define a logical storage unit, the file. Files are mapped by the operating system onto physical devices. These storage devices are usually nonvolatile, so the contents are persistent between system reboots.

A file is a named collection of related information that is recorded on secondary storage. From a user's perspective, a file is the smallest allotment of logical secondary storage; that is, data cannot be written to secondary storage unless they are within a file. Commonly, files represent programs（both source and object forms）and data. Data files may be numeric, alphabetic, alphanumeric, or binary. Files may be free form, such as text files, or may be formatted rigidly. In general, a file is a sequence of bits, bytes, lines, or records, the meaning of which is defined by the file's creator and user. The concept of a file is thus extremely general.

The information in a file is defined by its creator. Many different types of information may be stored in a file—source or executable programs, numeric or text data, photos, music, video, and so on. A file has a certain defined structure, which depends on its type. A text file is a sequence of characters organized into lines（and possibly pages）. A source file is a sequence of functions, each of which is further organized as declarations followed by executable statements. An executable file is a series of code sections that the loader can bring into memory and execute.

A file is an abstract data type defined and implemented by the operating system. It is a sequence of logical records. A logical record may be a byte, a line (of fixed or variable length), or a more complex data item. The operating system may specifically support various record types or may leave that support to the application program.

The major task for the operating system is to map the logical file concept onto physical storage devices such as magnetic disk or tape. Since the physical record size of the device may not be the same as the logical record size, it may be necessary to order logical records into physical records. Again, this task may be supported by the operating system or left for the application program.

Each device in a file system keeps a volume table of contents or a device directory listing the location of the files on the device. In addition, it is useful to create directories to allow files to be organized. A single-level directory in a multiuser system causes naming problems, since each file must have a unique name. A two-level directory solves this problem by creating a separate directory for each user's files. The directory lists the files by name and includes the file' s location on the disk, length, type, owner, time of creation, time of last use, and so on.

The natural generalization of a two-level directory is a tree-structured directory. A tree-structured directory allows a user to create subdirectories to organize files. Acyclic-graph directory structures enable users to share subdirectories and files but complicate searching and deletion. A general graph structure allows complete flexibility in the sharing of files and directories but sometimes requires garbage collection to recover unused disk space.

Disks are segmented into one or more volumes, each containing a file system or left "raw." File systems may be mounted into the system's naming structures to make them available. The naming scheme varies by operating system. Once mounted, the files within the volume are available for use. File systems may be unmounted to disable access or for maintenance.

File sharing depends on the semantics provided by the system. Files may have multiple readers, multiple writers, or limits on sharing. Distributed file systems allow client hosts to mount volumes or directories from servers, as long as they can access each other across a network. Remote file systems present challenges in reliability, performance, and security. Distributed information systems maintain user, host, and access information so that clients and servers can share state information to manage use and access.

Since files are the main information-storage mechanism in most computer systems, file protection is needed. Access to files can be controlled separately for each type of access—read, write, execute, append, delete, list directory, and so on. File protection can be provided by access lists, passwords, or other techniques.

The file system resides permanently on secondary storage, which is designed to hold a large amount of data permanently. The most common secondary-storage medium is the disk.

Physical disks may be segmented into partitions to control media use and to allow multiple, possibly varying, file systems on a single spindle. These file systems are mounted onto a logical file system architecture to make them available for use. File systems are often implemented in a layered or modular structure. The lower levels deal with the physical properties of storage devices. Upper levels deal with symbolic file names and logical properties of files. Intermediate levels map the logical file concepts into physical device properties.

Any file-system type can have different structures and algorithms. A VFS layer allows the upper layers to deal with each file-system type uniformly. Even remote file systems can be integrated into the system's directory structure and acted on by standard system calls via the VFS interface.

The various files can be allocated space on the disk in three ways: through contiguous, linked, or indexed allocation. Contiguous allocation can suffer from external fragmentation. Direct access is very inefficient with linked allocation. Indexed allocation may require substantial overhead for its index block. These algorithms can be optimized in many ways. Contiguous space can be enlarged through extents to increase flexibility and to decrease external fragmentation. Indexed allocation can be done in clusters of multiple blocks to increase throughput and to reduce the number of index entries needed. Indexing in large clusters is similar to contiguous allocation with extents.

Free-space allocation methods also influence the efficiency of disk-space use, the performance of the file system, and the reliability of secondary storage. The methods used include bit vectors and linked lists. Optimizations include grouping, counting, and the FAT, which places the linked list in one contiguous area.

Directory-management routines must consider efficiency, performance, and reliability. A hash table is a commonly used method, as it is fast and efficient. Unfortunately, damage to the table or a system crash can result in inconsistency between the directory information and the disk's contents. A consistency checker can be used to repair the damage. Operating-system backup tools allow disk data to be copied to tape, enabling the user to recover from data or even disk loss due to hardware failure, operating system bug, or user error.

Network file systems, such as NFS, use client-server methodology to allow users to access files and directories from remote machines as if they were on local file systems. System calls on the client are translated into network protocols and retranslated into file-system operations on the server. Networking and multiple-client access create challenges in the areas of data consistency and performance.

Due to the fundamental role that file systems play in system operation, their performance and reliability are crucial. Techniques such as log structures and caching help improve performance, while log structures and RAID improve reliability. The WAFL file system is an example of optimization of performance to match a specific I/O load.

9.7.2 Inode

The inode（index node）is a data structure in a Unix-style file system that describes a file-system object such as a file or a directory. Each inode stores the attributes and disk block locations of the object's data. File-system object attributes may include metadata（times of last change, access, modification）, as well as owner and permission data.

Directories are lists of names assigned to inodes. A directory contains an entry for itself, its parent, and each of its children.

A file system relies on data structures about the files, as opposed to the contents of that file. The former are called metadata—data that describes data. Each file is associated with an inode, which is identified by an integer, often referred to as an i-number or inode number.

Inodes store information about files and directories (folders), such as file ownership, access mode (read, write, execute permissions), and file type. On many older file system implementations, the maximum number of inodes is fixed at file system creation, limiting the maximum number of files the file system can hold. A typical allocation heuristic for inodes in a file system is one inode for every 2K bytes contained in the filesystem.

The inode number indexes a table of inodes in a known location on the device. From the inode number, the kernel's file system driver can access the inode contents, including the location of the file, thereby allowing access to the file. A file’s inode number can be found using the ls -i command. The ls -i command prints the i-node number in the first column of the report.

Some Unix-style file systems such as ReiserFS, btrfs, and APFS omit a fixed-size inode table, but must store equivalent data in order to provide equivalent capabilities. The data may be called stat data, in reference to the stat system call that provides the data to programs. Common alternatives to the fixed-size table include B-trees and the derived B+ trees.

File names and directory implications:

Inodes do not contain its hardlink names, only other file metadata.

Unix directories are lists of association structures, each of which contains one filename and one inode number.

The file system driver must search a directory looking for a particular filename and then convert the filename to the correct corresponding inode number.

The operating system kernel's in-memory representation of this data is called struct inode in Linux. Systems derived from BSD use the term vnode (the “v” refers to the kernel’s virtual file system layer).

9.7.3 Ext4

The ext4 journaling file system or fourth extended filesystem is a journaling file system for Linux, developed as the successor to ext3.

Ext4 was initially a series of backward-compatible extensions to ext3, many of them originally developed by Cluster File Systems for the Lustre file system between 2003 and 2006, meant to extend storage limits and add other performance improvements.However, other Linux kernel developers opposed accepting extensions to ext3 for stability reasons,and proposed to fork the source code of ext3, rename it as ext4, and perform all the development there, without affecting existing ext3 users. This proposal was accepted, and on 28 June 2006, Theodore Ts’o, the ext3 maintainer, announced the new plan of development for ext4.A preliminary development version of ext4 was included in version 2.6.19 of the Linux kernel. On 11 October 2008, the patches that mark ext4 as stable code were merged in the Linux 2.6.28 source code repositories,denoting the end of the development phase and recommending ext4 adoption. Kernel 2.6.28, containing the ext4 filesystem, was finally released on 25 December 2008.On 15 January 2010, Google announced that it would upgrade its storage infrastructure from ext2 to ext4. On 14 December 2010, Google also announced it would use ext4, instead of YAFFS, on Android 2.3.

Features

- Large file system

The ext4 filesystem can support volumes with sizes up to 1 exbibyte（EiB）and single files with sizes up to 16 tebibytes（TiB）with the standard 4 KiB block size. The maximum file, directory, and filesystem size limits grow at least proportionately with the filesystem block size up to the maximum 64 KiB block size available on ARM and PowerPC/Power ISA CPUs.

- Extents

Extents replace the traditional block mapping scheme used by ext2 and ext3. An extent is a range of contiguous physical blocks, improving large-file performance and reducing fragmentation. A single extent in ext4 can map up to 128 MiB of contiguous space with a 4 KiB block size.There can be four extents stored directly in the inode. When there are more than four extents to a file, the rest of the extents are indexed in a tree.

- Backward compatibility

Ext4 is backward-compatible with ext3 and ext2, making it possible to mount ext3 and ext2 as ext4. This will slightly improve performance, because certain new features of the ext4 implementation can also be used with ext3 and ext2, such as the new block allocation algorithm, without affecting the on-disk format.

Ext3 is partially forward-compatible with ext4. Practically, ext4 will not mount as an ext3 filesystem out of the box, unless certain new features are disabled when creating it.

- Persistent pre-allocation

Ext4 can pre-allocate on-disk space for a file. To do this on most file systems, zeroes would be written to the file when created. In ext4（and some other files systems such as XFS）fallocate(), a new system call in the Linux kernel, can be used. The allocated space would be guaranteed and likely contiguous. This situation has applications for media streaming and databases.

- Delayed allocation

Ext4 uses a performance technique called allocate-on-flush, also known as delayed allocation. That is, ext4 delays block allocation until data is flushed to disk.（In contrast, some file systems allocate blocks immediately, even when the data goes into a write cache.）Delayed allocation improves performance and reduces fragmentation by effectively allocating larger amounts of data at a time.

- Unlimited number of subdirectories

Ext4 does not limit the number of subdirectories in a single directory, except by the inherent size limit of the directory itself.（In ext3 a directory can have at most 32,000 subdirectories.）To allow for larger directories and continued performance, ext4 in Linux 2.6.23 and later turns on HTree indices（a specialized version of a B-tree）by default, which allows directories up to approximately 10-12 million entries to be stored in the 2-level HTree index and 2 GB directory size limit for 4 KiB block size, depending on the filename length. In Linux 4.12 and later the largedir feature enabled a 3-level HTree and directory sizes over 2 GB, allowing approximately 6 billion entries in a single directory.

- Journal checksums

Ext4 uses checksums in the journal to improve reliability, since the journal is one of the most used files of the disk. This feature has a side benefit: it can safely avoid a disk I/O wait during journaling, improving performance slightly. Journal checksumming was inspired by a research article from the University of Wisconsin, titled IRON File Systems (specifically, section 6, called "transaction checksums"), with modifications within the implementation of compound transactions performed by the IRON file system (originally proposed by Sam Naghshineh in the RedHat summit).

- Metadata checksumming

Since Linux kernel 3.5 released in 2012

- Faster file-system checking

In ext4 unallocated block groups and sections of the inode table are marked as such. This enables e2fsck to skip them entirely and greatly reduces the time it takes to check the file system. Linux 2.6.24 implements this feature.

- Multiblock allocator

When ext3 appends to a file, it calls the block allocator, once for each block. Consequently, if there are multiple concurrent writers, files can easily become fragmented on disk. However, ext4 uses delayed allocation, which allows it to buffer data and allocate groups of blocks. Consequently, the multiblock allocator can make better choices about allocating files contiguously on disk. The multiblock allocator can also be used when files are opened in O_DIRECT mode. This feature does not affect the disk format.

- Improved timestamps

As computers become faster in general, and as Linux becomes used more for mission-critical applications, the granularity of second-based timestamps becomes insufficient. To solve this, ext4 provides timestamps measured in nanoseconds. In addition, 2 bits of the expanded timestamp field are added to the most significant bits of the seconds field of the timestamps to defer the year 2038 problem for an additional 408 years.ext4 also adds support for time-of-creation timestamps. But, as Theodore Ts'o points out, while it is easy to add an extra creation-date field in the inode (thus technically enabling support for these timestamps in ext4), it is more difficult to modify or add the necessary system calls, like stat() (which would probably require a new version) and the various libraries that depend on them (like glibc). These changes will require coordination of many projects. Therefore, the creation date stored by ext4 is currently only available to user programs on Linux via the statx() API.

- Project quotas

Support for project quotas was added in Linux kernel 4.4 on 8 Jan 2016. This feature allows assigning disk quota limits to a particular project ID. The project ID of a file is a 32-bit number stored on each file and is inherited by all files and subdirectories created beneath a parent directory with an assigned project ID. This allows assigning quota limits to a particular subdirectory tree independent of file access permissions on the file, such as user and project quotas that are dependent on the UID and GID. While this is similar to a directory quota, the main difference is that the same project ID can be assigned to multiple top-level directories and is not strictly hierarchical.

- Transparent encryption

Support for transparent encryption was added in Linux kernel 4.1 on June 2015.

- Lazy initialization

The lazyinit feature allows to clean inode tables in background, speeding initialization when creating a new ext4 file system. It is available since 2010 in Linux kernel version 2.6.37.

- Write barriers

Ext4 enables write barriers by default. It ensures that file system metadata is correctly written and ordered on disk, even when write caches lose power. This goes with a performance cost especially for applications that use fsync heavily or create and delete many small files. For disks with a battery-backed write cache, disabling barriers（option 'barrier=0'）may safely improve performance.

Limitations

In 2008, the principal developer of the ext3 and ext4 file systems, Theodore Ts'o, stated that although ext4 has improved features, it is not a major advance, it uses old technology, and is a stop-gap. Ts'o believes that Btrfs is the better direction because "it offers improvements in scalability, reliability, and ease of management". Btrfs also has "a number of the same design ideas that reiser3/4 had". However, ext4 has continued to gain new features such as file encryption and metadata checksums.

The ext4 file system does not honor the "secure deletion" file attribute, which is supposed to cause overwriting of files upon deletion. A patch to implement secure deletion was proposed in 2011, but did not solve the problem of sensitive data ending up in the file-system journal.

9.8 实验7 Linux文件系统

获取视频

9.8.1 实验说明

本次实验从磁盘高级格式化开始，讲解分区及文件系统的装载等内容，特别针对 ext4 文件系统，深入介绍 inode 的工作原理。本实验的实验内容多，过程比较繁杂，请读者一定耐心完成，建议参照演示视频步骤完成。

9.8.2 磁盘高级格式化

1. 准备工作

(1) 转换成 root 身份。

```
$ su      #如果没有设置过 root 密码的，先使用 sudo passwd root 设置 root 密码
```

(2)查看当前系统的磁盘数据：

```
$ fdisk -l
Disk /dev/sda: 20 GiB, 21474836480 bytes, 41943040 sectors
Units: sectors of 1 * 512 = 512 bytes
Sector size (logical/physical): 512 bytes / 512 bytes
I/O size (minimum/optimal): 512 bytes / 512 bytes
Disklabel type: dos
Disk identifier: 0x6728fa32

Device     Boot Start     End  Sectors Size Id Type
/dev/sda1  *     2048 41943039 41940992  20G 83 Linux
```

(3)关闭虚拟机，在虚拟机控制台中增加一个磁盘，视频中以 VMWare 虚拟机为例，增加了一个 5GB 大小的磁盘。

(4)重启虚拟机，再利用 fdisk -l 指令观察数据变化。

```
Disk /dev/sdb: 5 GiB, 5368709120 bytes, 10485760 sectors
Units: sectors of 1 * 512 = 512 bytes
Sector size (logical/physical): 512 bytes / 512 bytes
I/O size (minimum/optimal): 512 bytes / 512 bytes
```

2. MBR 分区

(1)将 /dev/sdb 磁盘设备设置为 MBR 分区(disklabel type)，并创建一个分区。

```
$ fdisk /dev/sdb
Welcome to fdisk (util-linux 2.29.2).
Changes will remain in memory only, until you decide to write them.
Be careful before using the write command.

Command (m for help): o #设置该设备为 MBR 分区
Created a new DOS disklabel with disk identifier 0x87807b6a.

Command (m for help): p #打印出分区情况
Disk /dev/sdb: 5 GiB, 5368709120 bytes, 10485760 sectors
Units: sectors of 1 * 512 = 512 bytes
Sector size (logical/physical): 512 bytes / 512 bytes
I/O size (minimum/optimal): 512 bytes / 512 bytes
Disklabel type: dos
Disk identifier: 0x87807b6a

Command (m for help): n #增加一个分区
Partition type
   p   primary (0 primary, 0 extended, 4 free)
   e   extended (container for logical partitions)
Select (default p):
#以下全部采用了默认值，即将 5G 空间全部设置为 primary 主分区，分区号为 1
```

```
Using default response p.
Partition number (1-4, default 1):
First sector (2048-10485759, default 2048):
Last sector, +sectors or +size{K,M,G,T,P} (2048-10485759, default
10485759):

Created a new partition 1 of type 'Linux' and of size 5 GiB.

Command (m for help): p #重新输出sdb的分区情况
Disk /dev/sdb: 5 GiB, 5368709120 bytes, 10485760 sectors
Units: sectors of 1 * 512 = 512 bytes
Sector size (logical/physical): 512 bytes / 512 bytes
I/O size (minimum/optimal): 512 bytes / 512 bytes
Disklabel type: dos #MBR分区
Disk identifier: 0x7f1f925f #设备ID号
#分区号     起始扇区号     终止扇区号   总扇区数   容量   分区类型编号  分区类型
Device     Boot Start    End      Sectors   Size  Id     Type
/dev/sdb1 2048           10485759 10483712  5G    83     Linux

Command (m for help): w #用该指令应用上述分区修改，如果放弃，可以使用'q'指令
The partition table has been altered.
```


(2)分区的起始扇区号为2048，前面第0～2047号扇区为保留扇区，第0号扇区为MBR。

(3)观察/dev/sdb是否符合MBR的特征：MBR是磁盘驱动器上的第一个扇区。MBR包含引导程序代码(440字节)，可能还包含其他一些信息，紧接着是64字节的分区表和一个2字节的引导签名。64字节的分区表中有4个16字节的条目，从偏移量446 (1BEh) 开始。表9-2给出了每个16字节条目的布局。

表9-2　16字节条目的布局

偏移量（十六进制数）	长度	描述
0h	1	状态，80h表示活动（可引导）的分区
1h	3	分区中第一个绝对扇区的CHS（柱面-磁头-扇区）地址
4h	1	分区类型
5h	3	分区中最后一个绝对扇区的CHS地址
8h	4	分区中第一个绝对扇区的逻辑块地址（LBA）
ch	4	分区中的扇区数量

通过下面的指令，将MBR以十六进制数的形式打印出来。

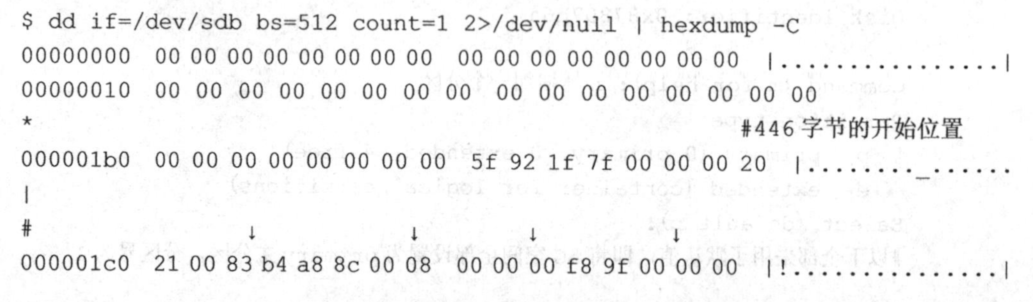

```
$ dd if=/dev/sdb bs=512 count=1 2>/dev/null | hexdump -C
00000000  00 00 00 00 00 00 00 00  00 00 00 00 00 00 00 00  |................|
00000010  00 00 00 00 00 00 00 00  00 00 00 00 00 00 00 00
*                                                 #446字节的开始位置
000001b0  00 00 00 00 00 00 00 00  5f 92 1f 7f 00 00 00 20  |........_......
|
#                    ↓          ↓        ↓           ↓
000001c0  21 00 83 b4 a8 8c 00 08  00 00 00 f8 9f 00 00 00  |!...............|
```

```
000001d0  00 00 00 00 00 00 00 00  00 00 00 00 00 00 00 00  |................|
*                                               #分区有效标志
000001f0  00 00 00 00 00 00 00 00  00 00 00 00 00 00 55 aa  |..............U.|
00000200
```

(4) 使用下面的指令只显示 64 字节的分区表信息。

```
$ dd if=/dev/sdb bs=510 count=1 2>/dev/null | tail -c 64 | hexdump -C
00000000  00 20 21 00 83 b4 a8 8c  00 08 00 00 00 f8 9f 00  |. !.............|
00000010  00 00 00 00 00 00 00 00  00 00 00 00 00 00 00 00  |................|
*
00000040
```

(5) 扇区 LBA 号使用 4 字节来编排，所以分区的最大扇区数为 2^{32} 个，扇区大小为 512 字节，则最大支持容量为 $2^{32}\times512=2$TB。.

(6) 分区数量最多是 4 个，为了超过这个限制，人们又提出扩展分区这个概念，在此不再展开，有兴趣的读者可自行研究。

3. GPT 分区

GPT（GUID Partition Table，GUID 分区表）分区方案是源自 EFI 标准的一种较新的磁盘分区表结构的标准。与普遍使用的主引导记录（MBR）分区方案相比，GPT 提供了更加灵活的磁盘分区机制。GPT 分区方案如图 9-16 所示。

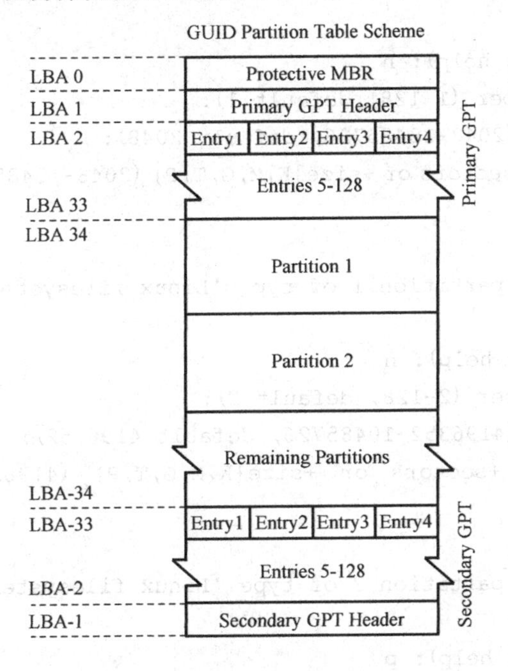

图 9-16　GPT 分区方案

(1) 扇区按 LBA 模式编排，第 0 号扇区存放一个叫 Protective MBR 的数据结构，是为了兼容 MBR 分区，里面的分区类型为 0xEE，不支持 GPT 的系统读到这个标志就会报错。

(2) 每个分区信息占用 128 字节，因此 1 个扇区可以存放 4 个分区信息，第 2～33 号扇区都是存放分区信息的，因此 GPT 最多可以支持 128 个分区。

(3)扇区号使用 8 字节进行编制，因此每个分区最大支持容量为 $2^{65}\times512$=8ZB(注意：B→KB→MB→GB→TB→PB→EB→ZB→YB→BB→NB→DB)

(4)使用 fdisk 将/dev/sdb 设备重新设置为 GPT 分区，并且按 2∶3 的比例划分出两个分区。

```
$ fdisk /dev/sdb

Welcome to fdisk (util-linux 2.29.2).
Changes will remain in memory only, until you decide to write them.
Be careful before using the write command.

Command (m for help): g
Created a new GPT disklabel (GUID: 7CE8EAC3-AC64-4EC6-BE11-711A5AF46A05).

Command (m for help): p
Disk /dev/sdb: 5 GiB, 5368709120 bytes, 10485760 sectors
Units: sectors of 1 * 512 = 512 bytes
Sector size (logical/physical): 512 bytes / 512 bytes
I/O size (minimum/optimal): 512 bytes / 512 bytes
Disklabel type: gpt
Disk identifier: 7CE8EAC3-AC64-4EC6-BE11-711A5AF46A05

Command (m for help): n
Partition number (1-128, default 1):
First sector (2048-10485726, default 2048):
Last sector, +sectors or +size{K,M,G,T,P} (2048-10485726, default 10485726):
+2G

Created a new partition 1 of type 'Linux filesystem' and of size 2 GiB.

Command (m for help): n
Partition number (2-128, default 2):
First sector (4196352-10485726, default 4196352):
Last sector, +sectors or +size{K,M,G,T,P} (4196352-10485726, default
10485726):

Created a new partition 2 of type 'Linux filesystem' and of size 3 GiB.

Command (m for help): p
Disk /dev/sdb: 5 GiB, 5368709120 bytes, 10485760 sectors
Units: sectors of 1 * 512 = 512 bytes
Sector size (logical/physical): 512 bytes / 512 bytes
I/O size (minimum/optimal): 512 bytes / 512 bytes
Disklabel type: gpt
Disk identifier: 7CE8EAC3-AC64-4EC6-BE11-711A5AF46A05
```

```
Device      Start     End Sectors Size Type
/dev/sdb1    2048  4196351 4194304   2G Linux filesystem
/dev/sdb2 4196352 10485726 6289375   3G Linux filesystem

Command (m for help): t
Partition number (1,2, default 2): 2
Hex code (type L to list all codes): L
  1 EFI System                     C12A7328-F81F-11D2-BA4B-00A0C93EC93B
  2 MBR partition scheme           024DEE41-33E7-11D3-9D69-0008C781F39F
  3 Intel Fast Flash               D3BFE2DE-3DAF-11DF-BA40-E3A556D89593
  4 BIOS boot                      21686148-6449-6E6F-744E-656564454649
  5 Sony boot partition            F4019732-066E-4E12-8273-346C5641494F
  6 Lenovo boot partition          BFBFAFE7-A34F-448A-9A5B-6213EB736C22
  7 PowerPC PReP boot              9E1A2D38-C612-4316-AA26-8B49521E5A8B
  8 ONIE boot                      7412F7D5-A156-4B13-81DC-867174929325
  9 ONIE config                    D4E6E2CD-4469-46F3-B5CB-1BFF57AFC149
 10 Microsoft reserved             E3C9E316-0B5C-4DB8-817D-F92DF00215AE
 11 Microsoft basic data           EBD0A0A2-B9E5-4433-87C0-68B6B72699C7
 12 Microsoft LDM metadata         5808C8AA-7E8F-42E0-85D2-E1E90434CFB3
 13 Microsoft LDM data             AF9B60A0-1431-4F62-BC68-3311714A69AD
 14 Windows recovery environment   DE94BBA4-06D1-4D40-A16A-BFD50179D6AC
 15 IBM General Parallel Fs        37AFFC90-EF7D-4E96-91C3-2D7AE055B174
 16 Microsoft Storage Spaces       E75CAF8F-F680-4CEE-AFA3-B001E56EFC2D
 17 HP-UX data                     75894C1E-3AEB-11D3-B7C1-7B03A0000000
 18 HP-UX service                  E2A1E728-32E3-11D6-A682-7B03A0000000
 19 Linux swap                     0657FD6D-A4AB-43C4-84E5-0933C84B4F4F
 20 Linux filesystem               0FC63DAF-8483-4772-8E79-3D69D8477DE4
 21 Linux server data              3B8F8425-20E0-4F3B-907F-1A25A76F98E8
 22 Linux root (x86)               44479540-F297-41B2-9AF7-D131D5F0458A
 23 Linux root (ARM)               69DAD710-2CE4-4E3C-B16C-21A1D49ABED3
 24 Linux root (x86-64)            4F68BCE3-E8CD-4DB1-96E7-FBCAF984B709
 25 Linux root (ARM-64)            B921B045-1DF0-41C3-AF44-4C6F280D3FAE
 26 Linux root (IA-64)             993D8D3D-F80E-4225-855A-9DAF8ED7EA97
 27 Linux reserved                 8DA63339-0007-60C0-C436-083AC8230908
 28 Linux home                     933AC7E1-2EB4-4F13-B844-0E14E2AEF915
 29 Linux RAID                     A19D880F-05FC-4D3B-A006-743F0F84911E
 30 Linux extended boot            BC13C2FF-59E6-4262-A352-B275FD6F7172
 31 Linux LVM                      E6D6D379-F507-44C2-A23C-238F2A3DF928
 32 FreeBSD data                   516E7CB4-6ECF-11D6-8FF8-00022D09712B
 33 FreeBSD boot                   83BD6B9D-7F41-11DC-BE0B-001560B84F0F

Hex code (type L to list all codes): 11
```

```
Changed type of partition 'Linux filesystem' to 'Microsoft basic data'.

Command (m for help): p
Disk /dev/sdb: 5 GiB, 5368709120 bytes, 10485760 sectors
Units: sectors of 1 * 512 = 512 bytes
Sector size (logical/physical): 512 bytes / 512 bytes
I/O size (minimum/optimal): 512 bytes / 512 bytes
Disklabel type: gpt
Disk identifier: 7CE8EAC3-AC64-4EC6-BE11-711A5AF46A05

Device       Start      End Sectors Size Type
/dev/sdb1     2048  4196351 4194304   2G Linux filesystem
/dev/sdb2  4196352 10485726 6289375   3G Microsoft basic data

Command (m for help): w
The partition table has been altered.
Calling ioctl() to re-read partition table.
Syncing disks.

root@youngyt-PC:/home/youngyt# fdisk -l
Disk /dev/sda: 20 GiB, 21474836480 bytes, 41943040 sectors
Units: sectors of 1 * 512 = 512 bytes
Sector size (logical/physical): 512 bytes / 512 bytes
I/O size (minimum/optimal): 512 bytes / 512 bytes
Disklabel type: dos
Disk identifier: 0x6728fa32

Device     Boot Start      End  Sectors Size Id Type
/dev/sda1  *     2048 41943039 41940992  20G 83 Linux

Disk /dev/sdb: 5 GiB, 5368709120 bytes, 10485760 sectors
Units: sectors of 1 * 512 = 512 bytes
Sector size (logical/physical): 512 bytes / 512 bytes
I/O size (minimum/optimal): 512 bytes / 512 bytes
Disklabel type: gpt
Disk identifier: 7CE8EAC3-AC64-4EC6-BE11-711A5AF46A05

Device       Start      End Sectors Size Type
/dev/sdb1     2048  4196351 4194304   2G Linux filesystem
/dev/sdb2  4196352 10485726 6289375   3G Microsoft basic data
#Done
```

(5)用 dd 指令验证一下 protective MBR 扇区中的 0xEE 标志，不支持 GPT 的系统读到这个标志会报错，支持 GPT 的系统就会知道这个分区用的是 GPT 而不是 MBR。

```
root@youngyt-PC:/mnt/linux# dd if=/dev/sdb bs=510 count=1 2>/dev/null |
tail -c 64 | hexdump -C
00000000  00 00 01 00 ee fe ff ff  01 00 00 00 ff ff 9f 00  |................|
00000010  00 00 00 00 00 00 00 00  00 00 00 00 00 00 00 00  |................|
*
00000040
```

4. 格式化

(1)目的：构建文件系统(磁盘高级格式化)。

(2)指令：mkfs。

```
root@youngyt-PC:/home/youngyt# mkfs.ext4 /dev/sdb1
mke2fs 1.43.4 (31-Jan-2017)
Creating filesystem with 524288 4k blocks and 131072 inodes
Filesystem UUID: c5648df2-5619-48b7-b7e3-3d8fcf7dae91
Superblock backups stored on blocks:
 32768, 98304, 163840, 229376, 294912

Allocating group tables: done
Writing inode tables: done
Creating journal (16384 blocks): done
Writing superblocks and filesystem accounting information: done

root@youngyt-PC:/home/youngyt# mkfs.ntfs /dev/sdb2
Cluster size has been automatically set to 4096 bytes.
Initializing device with zeroes: 100% - Done.
Creating NTFS volume structures.
mkntfs completed successfully. Have a nice day.
root@youngyt-PC:/home/youngyt#
```

5. 挂载分区

(1)目的：让分区可用。

(2)指令：mount。

```
root@youngyt-PC:/media# cd /mnt
root@youngyt-PC:/mnt# mkdir linux
root@youngyt-PC:/mnt# mkdir windows
root@youngyt-PC:/mnt# fdisk -l
Disk /dev/sda: 20 GiB, 21474836480 bytes, 41943040 sectors
Units: sectors of 1 * 512 = 512 bytes
Sector size (logical/physical): 512 bytes / 512 bytes
I/O size (minimum/optimal): 512 bytes / 512 bytes
Disklabel type: dos
```

```
Disk identifier: 0x6728fa32

Device     Boot Start      End  Sectors Size Id Type
/dev/sda1  *     2048 41943039 41940992  20G 83 Linux


Disk /dev/sdb: 5 GiB, 5368709120 bytes, 10485760 sectors
Units: sectors of 1 * 512 = 512 bytes
Sector size (logical/physical): 512 bytes / 512 bytes
I/O size (minimum/optimal): 512 bytes / 512 bytes
Disklabel type: gpt
Disk identifier: 7CE8EAC3-AC64-4EC6-BE11-711A5AF46A05

Device       Start      End Sectors Size Type
/dev/sdb1     2048  4196351 4194304   2G Linux filesystem
/dev/sdb2  4196352 10485726 6289375   3G Microsoft basic data
root@youngyt-PC:/mnt# mount /dev/sdb1 /mnt/linux/
root@youngyt-PC:/mnt# mount /dev/sdb2 /mnt/windows/
#Done
```


(3)查看分区挂载情况

```
root@youngyt-PC:/mnt/linux# lsblk -f
NAME   FSTYPE LABEL UUID                                 MOUNTPOINT
sda
└─sda1 ext4         c8993682-0699-4cbe-8688-58d73bbc49af /
sdb
├─sdb1 ext4         c5648df2-5619-48b7-b7e3-3d8fcf7dae91 /mnt/linux
└─sdb2 ntfs         24AEF58C71C9751B                     /mnt/windows
```

9.8.3 Linux 文件系统操作

1. 查看树形目录结构

```
root@youngyt-PC:/mnt/linux# tree
.
├── Lecture
│   ├── L01
│   └── L02
├── lost+found
└── test
```

2. 查看文件占用空间大小

```
root@youngyt-PC:/mnt/linux# ls -l
total 24
drwxr-xr-x 4 root root  4096 Apr  7 11:26 Lecture
drwx------ 2 root root 16384 Apr  7 11:13 lost+found
```

```
-rw-r--r-- 1 root root   12 Apr 7 11:20 test

root@youngyt-PC:/mnt/linux# cat test
hello world

root@youngyt-PC:/mnt/linux# du -h test
4.0K test
```

- test 文件大小只有 12 字节，但是占用了 4KB(8 个扇区)的磁盘空间。
- Cluster：簇。文件系统是以簇为单位进行空间分配的，簇的大小是可以调节的。

3. 查看文件的目录项

```
root@youngyt-PC:/mnt/linux# stat test
  File: test
#       实际大小          占用的扇区数量          簇的大小
  Size: 12             Blocks: 8             IO Block: 4096   regular file
#                      inode 编号              链接数
Device: 811h/2065d   inode: 12               Links: 1
#   ACL (644 是八进制数)                      用户编号                    用户组编号
Access: (0644/-rw-r--r--)  Uid: (    0/    root)   Gid: (    0/    root)
#        二进制数     110 100 100
#        八进制数      6   4   4
Access: 2020-04-07 11:25:04.477016832 +0800
Modify: 2020-04-07 11:20:47.387464472 +0800
Change: 2020-04-07 11:20:47.387464472 +0800
```

其中 inode 存放了文件的物理扇区位置，因为存放一个文件大多数需要若干扇区(簇)，为了保证目录大小的一致性，将这些占用的扇区号统一保存在一个 inode 结构中，每个文件都有一个对应的 inode，每个 inode 都有唯一的编号，所有的 inode 都存放在分区开始部分的一个叫 superblock(超级块)的地方，可以用 inode 编号在超级块中进行索引。

4. 使用 chmod 改变文件的 ACL

```
root@youngyt-PC:/mnt/linux# chmod 464  test
root@youngyt-PC:/mnt/linux# ls -l
total 8
drwxr-xr-x 4 root root 4096 Apr  7 11:26 Lecture
-r--rw-r-- 1 root root   12 Apr  7 11:20 test
root@youngyt-PC:/mnt/linux# chmod 777 test
root@youngyt-PC:/mnt/linux# ls -l
total 8
drwxr-xr-x 4 root root 4096 Apr  7 11:26 Lecture
-rwxrwxrwx 1 root root   12 Apr  7 11:20 test
```

5. 使用 debugfs 指令观察文件的扇区内容

```
root@youngyt-PC:/mnt/linux# debugfs /dev/sdb1
debugfs 1.43.4 (31-Jan-2017)
```

```
debugfs:  blocks test  #查看 test 文件占用的扇区号
33025
debugfs:  bdump 33025  #将指定编号的扇区打印出来
0000  6865 6c6c 6f20 776f 726c 640a 0000 0000  hello world.....
0020  0000 0000 0000 0000 0000 0000 0000 0000  ................
*

debugfs:  q  #退出 debugfs

#试着将 test 文件删除
root@youngyt-PC:/mnt/linux# rm test
root@youngyt-PC:/mnt/linux# ls -l
total 4
drwxr-xr-x 4 root root 4096 Apr  7 11:26 Lecture

#再次进入 debugfs 查看之前的扇区内容
root@youngyt-PC:/mnt/linux# debugfs /dev/sdb1
debugfs 1.43.4 (31-Jan-2017)
debugfs:  bdump 33025
#发现文件虽然删除了，但是扇区中的文件内容还在，可以被用来反删除
0000  6865 6c6c 6f20 776f 726c 640a 0000 0000  hello world.....
0020  0000 0000 0000 0000 0000 0000 0000 0000  ................
*
#Done
```

第 10 章　I/O 系统

获取视频

10.1　概述

在计算机系统中，I/O 设备数量、种类繁多，I/O 系统是用于实现数据输入/输出及存储的系统。操作系统进行设备管理的对象主要是 I/O 设备，设备管理的基本任务是完成用户发出的 I/O 请求，提高 I/O 操作速度及 I/O 设备的利用率。

I/O 系统可分为 I/O 硬件和 I/O 软件两部分，本章将分别讨论。首先介绍 I/O 硬件的基础知识，接着讨论三种 I/O 控制方式，然后讨论操作系统提供的 I/O 服务及这些服务的应用程序 I/O 接口的实现，最后介绍操作系统的内核 I/O 子系统，包括 I/O 调度、缓冲、缓存及假脱机。I/O 设备的发展是日新月异的，读者应该欣赏操作系统设计人员是如何从抽象的层面设计软硬件接口，以不变应万变，采用软硬件组合技术解决各类问题的。

10.2　I/O 硬件

10.2.1　硬件原理

图 10-1 是一个典型的计算机总线结构图，图中大部分硬件都属于 I/O 设备，它们还有一个共同的特征：所有 I/O 设备连接到控制器(或适配器)后，再通过总线与 CPU 和内存连接在一起。设备与计算机的连接点称为端口(Port)，如 USB 端口及串行端口等。总线(Bus)是一组线路及通过线路传输信息的协议，如 PCI 总线、串行总线及并行总线等。控制器(Controller)或适配器(Adapter)是一组电子部件，它和特定的设备是密切相关的。控制器是指一些简单的设备控制单元，通常集成在 I/O 设备或主板上，如串口控制器及 USB 控制器等。适配器通常因功能多而体积较大，会单独制作在电路板上，如显示适配器(俗称显卡)。

如图 10-2 所示，控制器内部有三类寄存器：控制寄存器、状态寄存器及数据寄存器。CPU 通过读/写这些寄存器来控制通信，对 I/O 设备的控制指令会写入控制寄存器中，数据寄存器是用于存放要写入设备的数据或是从设备中读出的数据，状态寄存器用来指示设备的各种状态，如任务是否完成、数据是否可读及设备是否有故障等。

一般来说，我们将 I/O 设备分成两类：块设备(Block Device)和字符设备(Character Device)。顾名思义，块设备的数据传输以块为单位，是指那些数据传输量较大的设备，如磁盘、光盘及 U 盘等；字符设备是指以字节为传输单位的设备，如键盘及显示器等。

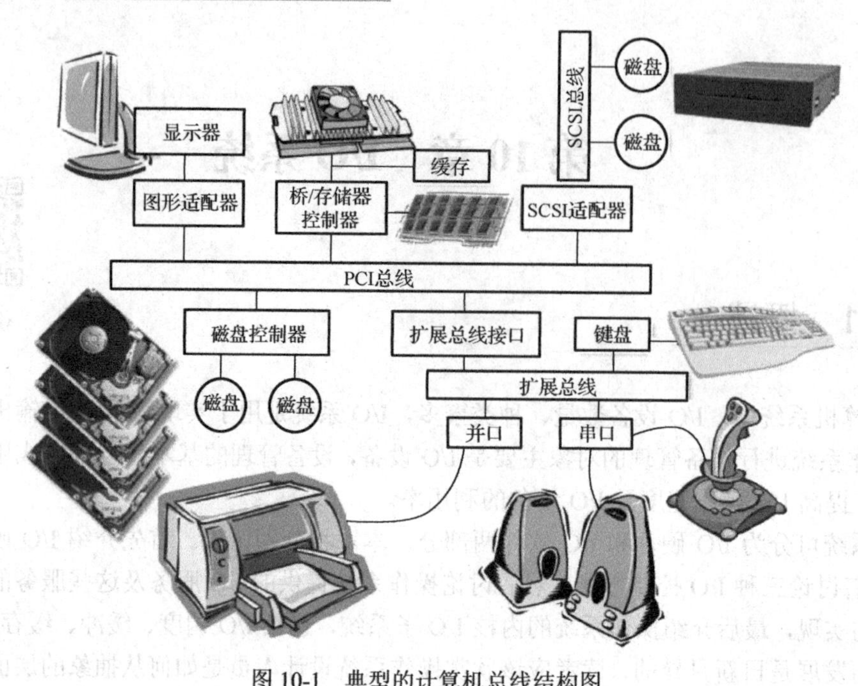

图 10-1　典型的计算机总线结构图

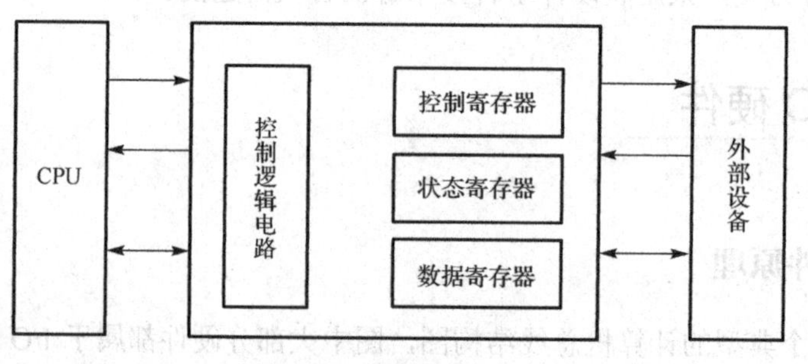

图 10-2　控制器原理图

I/O 控制在计算机处理中具有重要的地位，按控制器的功能强弱，以及和 CPU 之间联系方式的不同，我们介绍三种不同的 I/O 控制方式：轮询、中断及直接内存访问（Direct Memory Access，DMA）。

10.2.2　轮询

轮询（Polling）又称程序直接控制方式，主机与控制器之间采用握手的方式进行交互。下面看一个例子。

控制器的状态寄存器中的一个位［也叫忙位（busy bit）］用来表示设备状态（1 表示设备忙，0 表示设备空闲）；另一个位［就绪位（ready bit）］用来表示设备是否就绪（1 表示已就绪，0 表示未就绪）。双方的握手协议如下。

（1）主机重复读取 busy bit，直到该位为零（说明此时设备空闲）。

（2）主机设置控制寄存器的写位（表示要向设备进行写操作），并将要写的数据送到数据寄存器中。

（3）主机设置 ready bit 为 1，表示可以开始操作。

(4) 当控制器注意到 ready bit 已设置时，会设置 busy bit 为 1。

(5) 控制器读取控制寄存器，并看到写指令，它从数据寄存器中读取数据并向设备执行 I/O 写操作。

(6) 控制器清零 ready bit，清除状态寄存器的故障位，表示设备 I/O 操作成功，清零 busy bit，表示操作完成。

上述握手每次只能向设备写一字节数据，对于后续的每字节都重复这个循环。在步骤(1)中，主机处于忙等待(busy-waiting)或轮询状态，在该循环中，一直读取状态寄存器，直到 busy bit 被清零。如果控制器和设备都是快速设备，这段等待时间是可以接受的，但是如果等待时间过长，主机资源的利用率就会下降。于是，可以在此基础上进行改进，当设备处于忙状态时，可以让主机先做别的任务，当设备空闲后再通知主机。在现代计算机系统中，中断机制完美地实现了上述过程。

10.2.3　中断

在前面的章节中，我们讨论过的中断机制可以让设备具备向 CPU 反映自身状态的能力。图 10-3 给出了中断驱动的 I/O 循环，其过程如下。

(1) 主机(CPU) 发出设备初始化请求后不必查询设备状态，继续运行现有程序。

(2) I/O 控制器完成 I/O 请求的初始化工作。

(3) I/O 设备一旦就绪，就立刻产生一个中断。

(4) CPU 在执行指令期间都会进行中断检查，此时会捕获到设备就绪中断信号。

(5) 中断处理程序处理数据，从中断返回。

(6) 中断处理结束后，CPU 恢复被中断进程的状态。

(7) 循环进入一字节数据的传输请求。

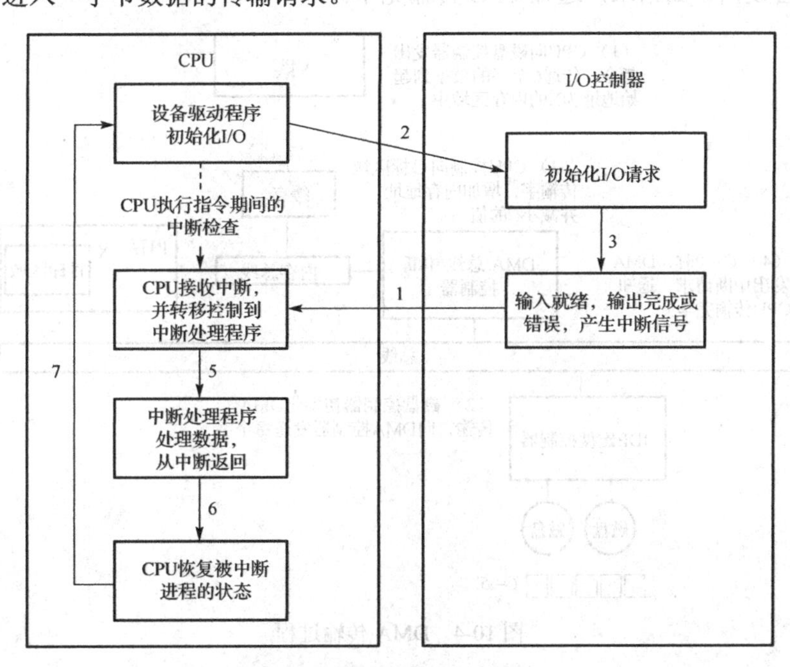

图 10-3　中断驱动的 I/O 循环

中断方式完美地解决了忙式等待问题，但是每次中断处理过程中传输的数据量仍然只有一字节，对于字符设备而言问题不大，但对于有大容量传输需求的块设备来讲，传输一个块需要启动太多次中断，消耗大量的CPU时间。我们都知道，无论是读设备还是写设备，CPU的工作是发布指令和接收结果，数据是在设备与内存之间流动，如果可以实现设备自主与内存交换数据，那么CPU就不必全程参与数据传输，在它们传输数据的时间内，CPU就可以为别的进程服务。当然这需要额外的硬件支持，也就是下面要介绍的DMA方式。

10.2.4 DMA

对于执行大量传输任务的设备(如磁盘驱动器)，如果通过昂贵的通用处理器来观察状态位并按字节来发送数据到控制器寄存器，似乎浪费了宝贵的资源。于是DMA出现了，它虽然也叫处理器，但它的任务相对简单，只负责控制数据在设备和内存之间的传输，因此它的造价很低。

DMA控制器至少需要以下逻辑部件：

(1)主存地址寄存器：要交换数据的内存起始地址；

(2)字计数器：传送数据的总字数，每传输一个字，该计数器减1；

(3)数据寄存器：暂存每次传送的数据。

一次完整的DMA传输过程如图10-4所示。若要从磁盘传输5个字到内存起始地址为X的区域，则CPU首先构造这个请求发送给磁盘控制器。DMA控制器接管传输工作，在没有CPU的帮助下启动这次传输，它会将指令告诉IDE磁盘控制器，磁盘控制器会从磁盘中读出这个5个字，依次发送给DMA控制器。DMA控制器每收到一个字就将其送到内存指定区域中，同时字计数器减1。当计数器为0时，说明数据传输完毕，DMA控制器的此次任务结束，DMA控制器会发出中断请求，通知CPU传输完毕。

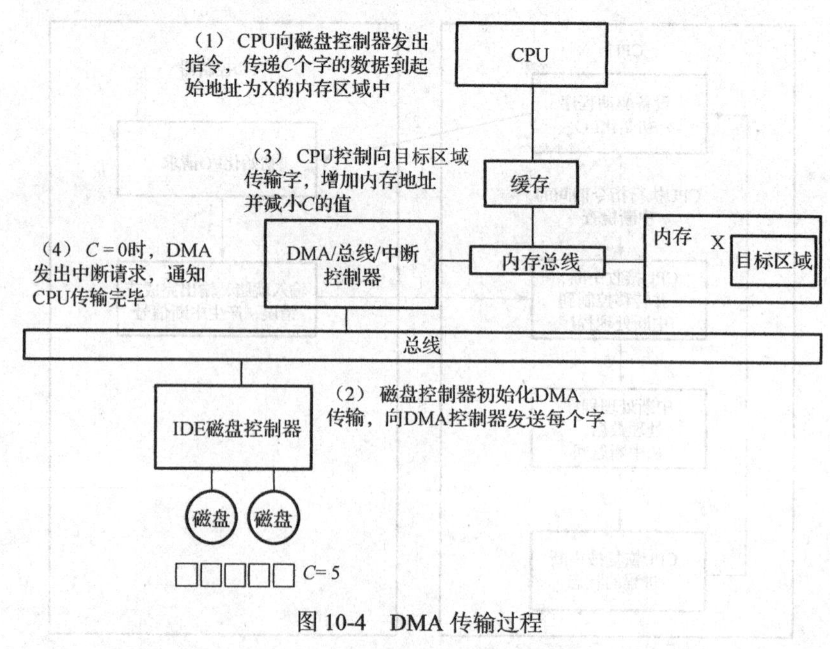

图10-4 DMA传输过程

在这个过程中，当DMA控制器占用内存总线时，CPU暂停访问内存，但是仍然可以访问主缓存或辅助缓存内的数据。这种 DMA 抢占 CPU 使用总线周期的现象叫周期窃取(Cycle Stealing)，这种行为当然会减慢 CPU 计算速度，但是如果不这么做，DMA 控制器可能很难得到内存总线来传输数据，何况 DMA 控制器占用内存总线的时间相对来说是比较短的。

目前，在小型、微型机中的快速设备均采用 DMA 控制方式。虽然 DMA 方式线路简单、价格低廉，但功能较差，对于复杂的 I/O 请求无能为力。所以在中大型机中一般使用通道方式，该方式的详细内容请阅读 10.5.2 节的阅读材料。

10.3　内核 I/O 结构

内核 I/O 结构包括 I/O 硬件和 I/O 软件两部分，10.2 节我们讨论了 I/O 硬件，本节开始介绍 I/O 软件部分。I/O 软件的总体设计目标是高效率(Efficiency)和通用性(Generality)。高效率旨在改善 I/O 设备的效率，特别是磁盘 I/O 的效率。通用性是指作为系统软件，在面对五花八门的硬件种类，每种设备不断更新换代时，能够做到“临危不乱”，不会疲于应付硬件的变化，而能够用统一的方法来管理所有设备，屏蔽硬件的细节，为用户使用硬件提供一个统一的接口。

内核 I/O 结构是一个层次结构，结构图如图 10-5 所示。在最底层是 I/O 硬件设备，与之直接连接的是设备控制器，这两层属于 I/O 硬件。与控制器相连的是设备驱动层(Device Driver Layer)，这层软件包括了与设备相关的代码，它的工作是把用户提交的逻辑 I/O 请求转化成物理 I/O 操作的启动和执行，该层的代码通常都是由设备的生产厂商提供的。再往上是内核 I/O 子系统(Kernel I/O Subsystem)，也叫“设备无关层”，该层软件不用考虑硬件的细节(这个工作是由设备驱动层完成的)，它的工作包括向设备驱动程序提供统一接口、设备命名、I/O 调度、缓冲、缓存、假脱机及错误报告等。

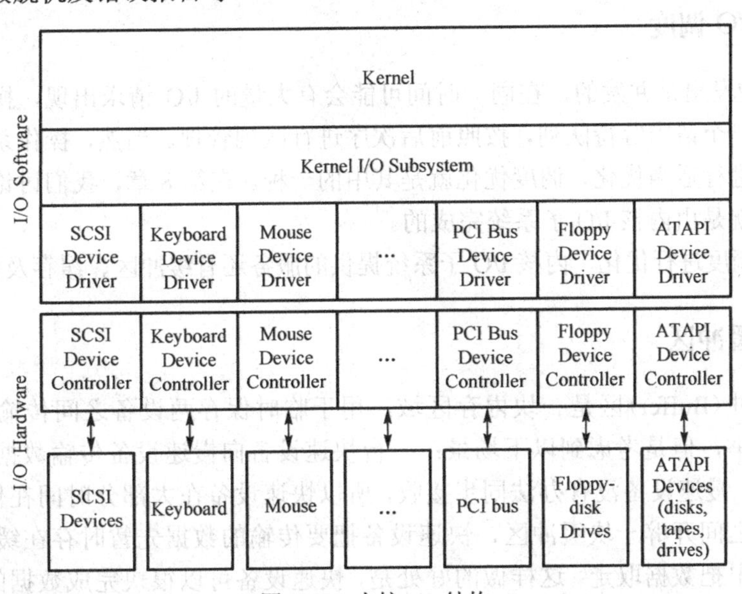

图 10-5　内核 I/O 结构

这种层次式的设计非常巧妙，在很多大型软件的开发场景中都能看到，为设备无关性打下了坚实的基础。试想一下，如果内核 I/O 子系统与驱动程序之间的接口是由生产厂商规定的，就会变成如图10-6(a)所示的样子，内核I/O子系统要适应不断出现的新硬件驱动。但如果由操作系统来规定接口规范，也就是图10-6(b)所示的形态，生产厂商按照统一的接口标准设计驱动程序，那么操作系统的数量就远远低于设备种类了。

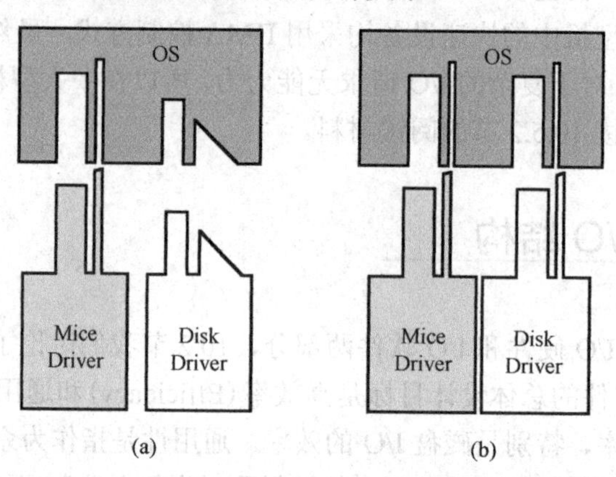

图10-6　内核I/O子系统和设备驱动层的接口设计

10.4 内核I/O子系统

内核I/O子系统非常重要，它起到上下衔接的作用，这一层所提供的服务和硬件本身是无关的，但是通过这一层的服务可以对I/O访问效率进行优化。本节将就内核I/O子系统的几个重要服务进行阐述。

10.4.1 I/O调度

I/O请求的发出是并发的，在同一时间可能会有大量的I/O请求出现，操作系统一般为每个设备维护一个请求等待队列，按照前后次序进行队列管理。当然，操作系统也可根据请求的设备种类进行适当优化，调度优化就是其中的一种。在第8章，我们讨论了磁盘的调度优化策略，其就是由内核I/O子系统完成的。

除了通过调度进行优化，内核I/O子系统提供的服务还有缓冲区、缓存及假脱机等。

10.4.2 缓冲区

当然，缓冲(Buffer)区是一块内存区域，用于临时保存两设备之间传输的数据。看上去好像多此一举，但是考虑到以下场景：一台快速设备向慢速设备传输数据，因为快速设备发数据太快，慢速设备没有办法同步接收，所以快速设备在大部分时间在做无用的等待。此时在两设备之间开辟一块缓冲区，快速设备把要传输的数据先暂时存在缓冲区中，慢速设备从缓冲区中把数据取走。这样做的好处是，快速设备可以很快完成数据的发送工作(虽然数据没有全部到达慢速设备而是在缓冲区中)转而去做其他工作，慢速设备也可以慢速读取缓冲区中的数据。在有些场景中，还可以设置两个缓冲区，甚至多个缓冲区。

缓冲区的用途是协调传输大小不一致数据的设备。这种不一致在计算机网络中特别常见，缓冲区大量用于消息的分段和重组。在发送端，一个大的消息被分成若干小的网络分组。这些网络分组通过网络被传输，而接收端先将它们放在重组缓冲区内，以便生成完整的源数据。

10.4.3 缓存

缓存(Cache)是保存数据副本的高速内存区域，访问缓存比访问原数据更加有效。例如，正在运行进程的指令保存在磁盘上，缓存在物理内存上，并被再次复制到 CPU 的次缓存和主缓存中。缓冲区和缓存的区别是，缓冲区可以保存数据项的唯一现有副本，而根据定义，缓存只是提供了一个位于其他地方的数据项的更快副本。

缓存和缓冲区的功能不同，但是有时一个内存区域可以用于两个目的。例如，为了保留复制语义和有效调度磁盘 I/O 操作，操作系统采用内存中的缓冲区来保存磁盘数据。这些缓冲区也作为缓存，以便提高文件的 I/O 效率；这些文件可被多个程序共享，或者快速地写入和重读。

当内核收到文件 I/O 请求时，内核首先访问缓冲区缓存，以便查看文件区域是否已经在内存中(可用)。如果是，可以避免或延迟物理磁盘 I/O 操作。此外，磁盘写入在数秒内会累积到缓冲区缓存中，以汇集大量传输来允许高效写入调度。

10.4.4 假脱机

假脱机(Simultaneous Peripheral Operations On-Line，SPOOLing)的意思是“外围设备联机操作”，它是关于慢速字符设备如何与计算机主机交换信息的一种技术。

有些设备(如打印机)是无法接收交叉的数据流的，每次只能打印一个任务，如果同时有多个任务到达，打印机是不可能进行混合输出的。操作系统采用的方案就是 SPOOLing 技术，系统会拦截打印作业，将打印输出到一个磁盘文件中，放置在 spool 区，如果有多个打印任务，那这些任务的打印输出文件在 spool 区中会按队列排序。系统会从 spool 区的打印队列中将输出文件一个一个依次送到打印机上执行真正的打印任务，整个过程由操作系统的系统进程 spooler 完成。操作系统还会提供一个控制界面(如图 10-7 所示)，以便用户和系统管理员查看队列，删除那些尚未打印而不再需要的任务，以及当打印机工作时暂停打印等。

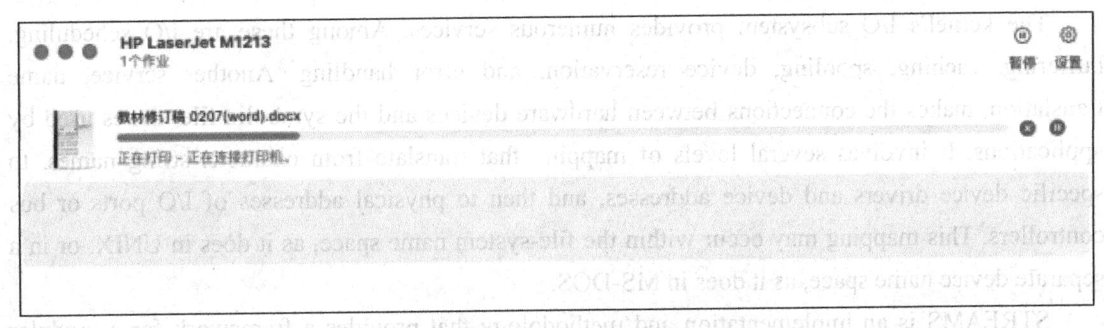

图 10-7 macOS 系统打印机 spooler 控制界面

10.5 Reading Materials

10.5.1 Overview

The control of devices connected to the computer is a major concern of operating-system designers. Because I/O devices vary so widely in their function and speed (consider a mouse, a hard disk, and a tape robot), varied methods are needed to control them. These methods form the I/O subsystem of the kernel, which separates the rest of the kernel from the complexities of managing I/O devices.

I/O-device technology exhibits two conflicting trends. On the one hand, we see increasing standardization of software and hardware interfaces. This trend helps us to incorporate improved device generations into existing computers and operating systems. On the other hand, we see an increasingly broad variety of I/O devices. Some new devices are so unlike previous devices that it is a challenge to incorporate them into our computers and operating systems. This challenge is met by a combination of hardware and software techniques. The basic I/O hardware elements, such as ports, buses, and device controllers, accommodate a wide variety of I/O devices. To encapsulate the details and oddities of different devices, the kernel of an operating system is structured to use device-driver modules. The device drivers present a uniform device-access interface to the I/O subsystem, much as system calls provide a standard interface between the application and the operating system.

The basic hardware elements involved in I/O are buses, device controllers, and the devices themselves. The work of moving data between devices and main memory is performed by the CPU as programmed I/O or is offloaded to a DMA controller. The kernel module that controls a device is a device driver. The system-call interface provided to applications is designed to handle several basic categories of hardware, including block devices, character devices, memory-mapped files, network sockets, and programmed interval timers. The system calls usually block the processes that issue them, but nonblocking and asynchronous calls are used by the kernel itself and by applications that must not sleep while waiting for an I/O operation to complete.

The kernel's I/O subsystem provides numerous services. Among these are I/O scheduling, buffering, caching, spooling, device reservation, and error handling. Another service, name translation, makes the connections between hardware devices and the symbolic file names used by applications. It involves several levels of mapping that translate from character-string names, to specific device drivers and device addresses, and then to physical addresses of I/O ports or bus controllers. This mapping may occur within the file-system name space, as it does in UNIX, or in a separate device name space, as it does in MS-DOS.

STREAMS is an implementation and methodology that provides a framework for a modular and incremental approach to writing device drivers and network protocols. Through streams, drivers can be stacked, with data passing through them sequentially and bidirectionally for processing.

I/O system calls are costly in terms of CPU consumption because of the many layers of software between a physical device and an application. These layers imply overhead from several sources: context switching to cross the kernel's protection boundary, signal and interrupt handling to service the I/O devices, and the load on the CPU and memory system to copy data between kernel buffers and application space.

10.5.2 I/O Channel

Many I/O tasks can be complex and require logic to be applied to the data to convert formats and other similar duties. In these situations, the simplest solution is to ask the CPU to handle the logic, but because I/O devices are relatively slow, a CPU could waste time (in computer perspective) waiting for the data from the device. This situation is called 'I/O bound'.

Channel architecture avoids this problem by using a logically independent, low-cost facility. Channel processors are simple, but self-contained, with minimal logic and sufficient scratchpad memory (working storage) to handle I/O tasks. They are typically not powerful or flexible enough to be used as a computer on their own and can be construed as a form of coprocessor. On some systems the channels use memory or registers addressable by the central processor as their scratchpad memory, while on other systems it is present in the channel hardware.

A CPU designates a block of storage or sends a relatively small channel programs to the channel in order to handle I/O tasks, which the channel and controller can, in many cases, complete without further intervention from the CPU (exception: those channel programs which utilize 'program controlled interrupts', PCIs, to facilitate program loading, demand paging and other essential system tasks).

When I/O transfer is complete or an error is detected, the controller communicates with the CPU through the channel using an interrupt. Since the channel has direct access to the main memory, it is also often referred to as a direct memory access (DMA) controller.

In the most recent implementations, the channel program is initiated and the channel processor performs all required processing until either an ending condition. This eliminates much of the CPU—Channel interaction and greatly improves overall system performance. The channel may report several different types of ending conditions, which may be unambiguously normal, may unambiguously indicate an error or whose meaning may depend on the context and the results of a subsequent sense operation. In some systems an I/O controller can request an automatic retry of some operations without CPU intervention. In earlier implementations, any error, no matter how small, required CPU intervention, and the overhead was, consequently, much higher. A program-controlled interruption (PCI) is still supported for certain "legacy" operations, but the trend is to move away from such PCIs, except where unavoidable.

A channel is an independent hardware component that co-ordinate all I/O to a set of controllers. Computer systems that use I/O channel have special hardware components that handle all I/O operations.

Channels use separate, independent and low cost processors for its functioning which are called Channel Processors.

Channel processors are simple, but contains sufficient memory to handle all I/O tasks. When I/O transfer is complete or an error is detected, the channel controller communicates with the CPU using an interrupt, and informs CPU about the error or the task completion.

Each channel supports one or more controllers or devices. Channel programs contain list of commands to the channel itself and for various connected controllers or devices. Once the operating system has prepared a list of I/O commands, it executes a single I/O machine instruction to initiate the channel program, the channel then assumes control of the I/O operations until they are completed.

A channel program is a sequence of channel command words（CCWs）that are executed by the I/O channel subsystem in the IBM System/360 and subsequent architectures. A channel program consists of one or more channel command words. The operating system signals the I/O channel subsystem to begin executing the channel program with an SSCH（start sub-channel）instruction. The central processor is then free to proceed with non-I/O instructions until interrupted. When the channel operations are complete, the channel interrupts the central processor with an I/O interruption. In earlier models of the IBM mainframe line, the channel unit was an identifiable component, one for each channel. In modern mainframes, the channels are implemented using an independent RISC processor, the channel processor, one for all channels. IBM System/370 Extended Architecture[3] and its successors replaced the earlier SIO（start I/O）and SIOF（start I/O fast release）machine instructions（System/360 and early System/370）with the SSCH（start sub-channel）instruction（ESA/370 and successors）.

Channel I/O provides considerable economies in input/output. For example, on IBM's Linux on IBM Z, the formatting of an entire track of a DASD requires only one channel program（and thus only one I/O instruction）, but multiple channel command words（one per block）. The program is executed by the dedicated I/O processor, while the application processor（the CPU）is free for other work.

10.5.3 The Buffer Cache

Reading from a disk is very slow compared to accessing（real）memory. In addition, it is common to read the same part of a disk several times during relatively short periods of time. For example, one might first read an e-mail message, then read the letter into an editor when replying to it, then make the mail program read it again when copying it to a folder. Or, consider how often the command ls might be run on a system with many users. By reading the information from disk only once and then keeping it in memory until no longer needed, one can speed up all but the first read. This is called disk buffering, and the memory used for the purpose is called the buffer cache.

Since memory is, unfortunately, a finite, nay, scarce resource, the buffer cache usually cannot be big enough（it can't hold all the data one ever wants to use）. When the cache fills up, the data that has been unused for the longest time is discarded and the memory thus freed is used for the new data.

Disk buffering works for writes as well. On the one hand, data that is written is often soon read again (e.g., a source code file is saved to a file, then read by the compiler), so putting data that is written in the cache is a good idea. On the other hand, by only putting the data into the cache, not writing it to disk at once, the program that writes runs quicker. The writes can then be done in the background, without slowing down the other programs.

Most operating systems have buffer caches (although they might be called something else), but not all of them work according to the above principles. Some are write-through: the data is written to disk at once (it is kept in the cache as well, of course). The cache is called write-back if the writes are done at a later time. Write-back is more efficient than write-through, but also a bit more prone to errors: if the machine crashes, or the power is cut at a bad moment, or the floppy is removed from the disk drive before the data in the cache waiting to be written gets written, the changes in the cache are usually lost. This might even mean that the filesystem (if there is one) is not in full working order, perhaps because the unwritten data held important changes to the bookkeeping information.

Because of this, you should never turn off the power without using a proper shutdown procedure or remove a floppy from the disk drive until it has been unmounted (if it was mounted) or after whatever program is using it has signaled that it is finished and the floppy drive light doesn't shine anymore. The sync command flushes the buffer, i.e., forces all unwritten data to be written to disk, and can be used when one wants to be sure that everything is safely written. In traditional UNIX systems, there is a program called update running in the background which does a sync every 30 seconds, so it is usually not necessary to use sync. Linux has an additional daemon, bdflush, which does a more imperfect sync more frequently to avoid the sudden freeze due to heavy disk I/O that sync sometimes causes.

Under Linux, bdflush is started by update. There is usually no reason to worry about it, but if bdflush happens to die for some reason, the kernel will warn about this, and you should start it by hand (/sbin/update).

The cache does not actually buffer files, but blocks, which are the smallest units of disk I/O (under Linux, they are usually 1 KB). This way, also directories, super blocks, other filesystem bookkeeping data, and non-filesystem disks are cached.

The effectiveness of a cache is primarily decided by its size. A small cache is next to useless: it will hold so little data that all cached data is flushed from the cache before it is reused. The critical size depends on how much data is read and written, and how often the same data is accessed. The only way to know is to experiment.

If the cache is of a fixed size, it is not very good to have it too big, either, because that might make the free memory too small and cause swapping (which is also slow). To make the most efficient use of real memory, Linux automatically uses all free RAM for buffer cache, but also automatically makes the cache smaller when programs need more memory.

Under Linux, you do not need to do anything to make use of the cache, it happens completely automatically. Except for following the proper procedures for shutdown and removing floppies, you do not need to worry about it.

10.6 实验8 Linux驱动实验

1. 实验目的

掌握Linux中module模块的编程基本方法，并可以将其动态载入内核。

2. 实验内容

(1) 创建一个空的C文件：hello.c，将下面的代码复制进文件。

```
#include <linux/kernel.h>
#include <linux/module.h>
#include <linux/init.h>
static int __init init_my_module(void){
  printk(KERN_INFO "Hello, my module!\n");
  return 0;
}
static void __exit exit_my_module(void){
  printk(KERN_INFO "Bye, my module!\n");
}
module_init(init_my_module);
module_exit(exit_my_module);
MODULE_LICENSE("GPL");
MODULE_AUTHOR("MyName");
```


(2) 在同一目录下创建一个空的文件：Makefile，将下面的代码复制进去。

注意：第3行和第5行的前面一定不能用空格，一定要用Tab键增加1个缩进。

```
obj-m += hello.o
all:
	make -C /lib/modules/$(shell uname -r)/build M=$(PWD) modules
clean:
	make -C /lib/modules/$(shell uname -r)/build M=$(PWD) clean
```

(3) 执行指令make，如果成功，应该在目录中生成一个hello.ko的内核模块。

(4) 使用指令insmod hello.ko，将模块动态插入内核，若提示权限问题，则在指令前面加上sudo。

(5) 执行指令dmesg，查看输出中是否有模块初始化打印的信息。

(6) 使用指令rmmod hello将模块动态地从内核中卸载，然后用dmesg指令查看内核的输出信息。

参 考 文 献

[1] Silberschatz A. Operating System Concepts [M]. 9th ed: Joh Wiley & Sons, 2013.

[2] 汤小丹，等.计算机操作系统[M]. 第 3 版. 西安：西安电子科技大学出版社，2007.

[3] 陈莉君，等. Linux 操作系统原理与应用[M]. 第 2 版. 北京：清华大学出版社，2012.

[4] 费翔林，等.操作系统教程[M]. 第 5 版. 北京：高等教育出版社，2014.

[5] Silberschatz A 著，郑扣根，译. 操作系统概念[M]. 第 9 版. 北京：机械工业出版社，2018.